Schwann Cell Development and Pathology

Kazunori Sango • Junji Yamauchi
Editors

Schwann Cell Development and Pathology

Springer

Editors
Kazunori Sango
Department of Sensory and Motor Systems
Tokyo Metropolitan Institute
of Medical Science
Tokyo, Japan

Junji Yamauchi
Department of Pharmacology
National Research Institute
for Child Health and Development
Tokyo, Japan

ISBN 978-4-431-54763-1 ISBN 978-4-431-54764-8 (eBook)
DOI 10.1007/978-4-431-54764-8
Springer Tokyo Heidelberg New York Dordrecht London

Library of Congress Control Number: 2013956818

Printed on acid-free paper

Springer is part of Springer Science+Business Media (www.springer.com)

Foreword

Peripheral nerve regeneration with functional recovery after injury, as well as the cure of various neuropathies, has been anticipated but remains to be achieved. For these accomplishments to occur, it is essential to elucidate the mechanisms and pathologies that are involved. One of the keys to clarifying the related problems is the Schwann cell, which is a type of peripheral nerve glial cell that surrounds and interacts with axons. Although myelinating Schwann cells form myelin around fast-conducting, large-diameter axons, nonmyelinating Schwann cells surround smaller-diameter axons without forming a myelin sheath. Schwann cells play essential roles in functions and structure in peripheral nerve from developmental to mature stages.

The rapid progress of molecular biological techniques in past decades, especially for RNA techniques and gene modification technologies, has brought new insights into the pathobiology of Schwann cells in vivo and in vitro. Studies combining recent stem cell biology with recent biotechnology, which is now closely linked to physicochemical fields, further explain how Schwann cell lineages develop, a process that has long been thought to be very complicated in vivo. These findings contribute to the elucidation of fundamental mechanisms during development and under pathological conditions.

Schwann Cell Development and Pathology presents recent topics in the development, differentiation, and myelination of Schwann cells, as well as pathological mechanisms and therapeutic approaches for peripheral neuropathies such as Charcot-Marie-Tooth diseases, amyloid polyneuropathy, immune-mediated neuropathy, and diabetic neuropathy. This book is certain to arouse the interest of readers in Schwann cell biology to meet the challenge of neural regeneration as well as the cure of neuropathies in the peripheral nervous system.

Yokohama, Japan　　　　　　　　　　Hidenori Horie

Preface

Schwann cells, named in honor of the German scientist Theodore Schwann (1810–1882), are recognized as glial cells in the peripheral nervous system and play major roles in development, maturation, and axonal regeneration and remyelination after injury. During the development and regeneration of peripheral nerves such as sciatic nerves, Schwann cells proliferate and migrate along the axons to their final destinations, where they eventually wrap around individual axons to form the myelin sheaths. Over time, myelin sheaths can grow to be more than 100 times larger than the collective surface area of premyelinating Schwann cell plasma membranes. The myelin sheath insulates axons to increase the nerve conduction velocity, which is termed saltatory conduction. It also protects axons from various physical stresses. In addition to the myelinating Schwann cells, Schwann cells are further categorized into three groups: nonmyelinating Schwann cells, perisynaptic Schwann cells, and satellite cells. All these cells are derived from neural crest cells. Although myelination by the Schwann cell is the event that occurs in both large-diameter ($A\alpha/\beta$) and small-diameter ($A\delta$) axons, the nonmyelinating Schwann cell can surround the small-diameter (C) axons, which originate from peripheral ganglia and consist of sympathetic and sensory neurons. The perisynaptic Schwann cell structurally and functionally helps to bridge the nerve terminal with its peripheral tissue to form the triparticle structures such as neuromuscular junctions. The satellite cell primarily associates with neuronal cell bodies positioned in the peripheral ganglia and appears to play a role in separating the respective cell body units. Considering these key functions of Schwann cells in the development and homeostasis of the peripheral nervous system, it is likely that the abnormalities of Schwann cells and their cross-talk with neurons lead to various critical peripheral nerve disorders, such as Charcot-Marie-Tooth diseases, amyloid polyneuropathy, immune-mediated neuropathy, and diabetic neuropathy.

The rapid progress of molecular biological techniques in past decades, especially for RNA techniques and gene modification technologies, has allowed us to investigate the pathobiology of Schwann cells in vivo and in vitro. Studies combining recent stem cell biology with recent biotechnology, which is now closely linked to physicochemical fields, further explain how Schwann cell lineages develop, a process that has long been thought to be very complicated in vivo. The findings contribute to the elucidation of fundamental mechanisms during development and under pathological conditions. We now know that these are closely tied to each other.

Despite such biological significance, neuroscientists and neurologists have paid less attention to Schwann cells than to glial cells in the central nervous system, and very few technical books on Schwann cells are currently available. This book presents recent topics on development, differentiation, and myelination of Schwann cells, as well as pathological mechanisms and therapeutic approaches for the peripheral neuropathies just described. The book also introduces unique co-culture systems to reproduce the neuron–Schwann cell interplay during development, degeneration and regeneration. As an original contribution, we fully describe spontaneously immortalized Schwann cell lines from normal adult mice (IMS32) and rats (IFRS1), and murine models of neurodegenerative disorders (e.g., lysosomal storage diseases, neurofibromatosis, and CMT1B). These cell lines are valuable tools for exploring neuron–Schwann cell interactions, the pathobiology of axonal degeneration and regeneration, and novel therapeutic approaches against neurological disorders in patients with relevant diseases.

We are confident that up-to-date research topics with high-quality immunofluorescence and electron micrographs introduced by young and energetic contributors will arouse the interest of readers in Schwann cell biology. Discussion from the point of view of basic and clinical neuroscience makes the book educational for medical students and young clinicians. As editors, we give thanks to all the authors and dedicate this book to the late Kyoko Ajiki, who continuously provided us excellent light and electron micrographs at the Tokyo Metropolitan Institute for Neuroscience but unfortunately died of cancer in 2011.

Tokyo, Japan

Kazunori Sango
Junji Yamauchi

Contents

1 **Introduction** 1
Kazunori Sango and Junji Yamauchi

2 **Recent Insights into Molecular Mechanisms That Control Growth Factor Receptor-Mediated Schwann Cell Morphological Changes During Development** 5
Yuki Miyamoto and Junji Yamauchi

3 **Membrane Skeleton in Schmidt–Lanterman Incisure in Schwann Cells of the Peripheral Nervous System** 29
Nobuo Terada, Yurika Saitoh, Nobuhiko Ohno, and Shinichi Ohno

4 **Schwann Cell–Axon Interactions: The Molecular and Metabolic Link Between Schwann Cells and Axons** 47
Nobuhiko Ohno, Takashi Sakoh, Yurika Saitoh, Nobuo Terada, and Shinichi Ohno

5 **Schwann Cell-Dependent Regulation of Peripheral Nerve Injury and Repair** 69
Keiichiro Susuki

6 **Charcot-Marie-Tooth Disease** 81
Toru Ogata

7 **Expression of the Transthyretin Gene in Schwann Cells and Familial Amyloidotic Polyneuropathy-Mediated Neurodegeneration** 103
Tatsufumi Murakami and Yoshihide Sunada

8 **Node of Ranvier Disruption: A Key Pathophysiology in Immune-Mediated Neuropathies** 121
Keiichiro Susuki

9 Pathogenesis of Diabetic Neuropathy from the Point of View of Schwann Cell Abnormalities 135
Koichi Kato, Eva L. Feldman, and Jiro Nakamura

10 Spontaneously Immortalized Adult Rodent Schwann Cells as Valuable Tools for the Study of Peripheral Nerve Degeneration and Regeneration 147
Kazunori Sango, Masami Tsukamoto, Kazunori Utsunomiya, and Kazuhiko Watabe

Index 171

Chapter 1
Introduction

Kazunori Sango and Junji Yamauchi

Abstract Schwann cells have key roles in development, differentiation, physiological homeostasis, and axonal regeneration and remyelination after injury in the peripheral nervous system. The abnormalities of Schwann cells and their crosstalk with neurons lead to peripheral nerve disorders, such as Charcot-Marie-Tooth disease, immune-mediated neuropathy, amyloid polyneuropathy, and diabetic neuropathy. In this book, we summarize recent topics on the biological features of Schwann cells under normal and pathological conditions, and introduce spontaneously immortalized Schwann cell lines from normal adult rodents and murine models of neurodegenerative disorders.

Keywords Development and maturation • Molecular biology • Myelination • Pathophysiology • Peripheral neuropathies • Schwann cells

1.1 Overview of the Book

An increasing number of reports have contributed to the elucidation of how Schwann cell lineages differentiate and develop, especially for myelination processes, leading to their unique three-dimensional structure. Recent evidence clarifies how cell-surface receptors, such as growth factor receptors, are associated with intracellular molecules, resulting in upregulation of myelin-related gene products and RNAs.

K. Sango (✉)
Laboratory of Peripheral Nerve Pathophysiology (ALS/Neuropathy Project), Department of Sensory and Motor Systems, Tokyo Metropolitan Institute of Medical Science, 2-1-6 Kamikitazawa, Setagaya-ku, Tokyo 156-8506, Japan
e-mail: sango-kz@igakuken.or.jp

J. Yamauchi
Department of Pharmacology, National Research Institute for Child Health and Development, 2-10-1 Okura, Setagaya, Tokyo 157-8535, Japan

K. Sango and J. Yamauchi (eds.), *Schwann Cell Development and Pathology*,
DOI 10.1007/978-4-431-54764-8_1,

Importantly, these findings are now being linked to mechanisms for the development and progression of critical neurological disorders, including Charcot-Marie-Tooth disease, amyloid polyneuropathy, immune-mediated neuropathy, and diabetic neuropathy. The rapid biotechnological progress in the past decades allows us to investigate the pathobiology of Schwann cells. The technologies uniquely combine co-culture systems to reproduce the neuron–Schwann cell interplay with analyzing Schwann cell degeneration and regeneration. We describe recent advances in the knowledge of mechanisms that underlie myelination and demyelination during development and pathology in Schwann cells.

This book consists of ten chapters, including this chapter. The biological properties of Schwann cells during development, differentiation, physiological homeostasis, and axonal degeneration and regeneration are illustrated in Chaps. 2, 3, 4, and 5, and Schwann cell abnormalities in various peripheral neuropathies are discussed in Chaps. 6, 7, 8, and 9. In addition, the originality of this book can be attributed, at least partly, to introducing basic features of spontaneously immortalized Schwann cell lines from adult rodents in Chap. 10.

1.2 Biological Properties of Schwann Cells During Development, Maturation, and Axonal Degeneration and Regeneration (Chaps. 2, 3, 4, and 5)

Schwann cell myelination processes are categorized into three stages: cell migration, cell elongation along axons, and axonal myelination, although these stages are not independent of each other but are a series of continuous processes. In Chap. 2, Miyamoto and Yamauchi discuss recent progress in unique, intracellular molecular mechanisms focusing on growth factor receptor-mediated Schwann cell morphological changes throughout the myelination processes, possibly identifying one basic mechanism of forming Schwann cell morphology (Yamauchi et al. 2012; Torii et al. 2013). The Schmidt–Lanterman incisure, a truncated cone shape in a myelin internode, is observed in the mature myelin structure. This structure is a specific feature of myelinated nerve fibers in the peripheral nervous systems. Terada et al. discuss a novel Src–MPP6–4.1G–CADM4 membrane skeletal molecular complex in Schmidt–Lanterman incisures, with potential roles in regulation of adhesion and signal transduction in Schwann cells (Chap. 3). Evidence suggests that the trophic support of Schwann cells is associated with modulation of axonal metabolism, which is involved in functional maintenance of axonal mitochondria. Ohno et al. discuss the recent molecular and cellular mechanisms of Schwann cell–axon interactions, helping us to understand how human pathological states arise and to develop new potential therapies of peripheral nervous system diseases (Chap. 4). Importantly, Schwann cells have the ability to trans-differentiate to distinct phenotypes after nerve injury and contribute to myelin debris clearance, attract macrophages into the lesions, protect injured neurons, promote axonal regrowth, and finally remyelinate and reform nodes of Ranvier along the regenerated axons. Susuki discusses recent knowledge concerning the unique Schwann cell plasticity (Chap. 5).

1.3 Schwann Cell Abnormalities in the Pathogenesis of Peripheral Neuropathies (Chaps. 6, 7, 8, and 9)

Ogata reviews recent progress in pathogenetic mechanisms with a focus on the causative genes and therapeutic approaches for Charcot-Marie-Tooth disease (Chap. 6). It is without doubt that the novel ideas reviewed in this chapter help us better understand how myelination is regulated and maintained in peripheral nerves. Murakami and Sunada have raised an intriguing hypothesis that transthyretin (TTR) produced in Schwann cells can be a cause of neurodegeneration in familial amyloid polyneuropathy (Chap. 7). On the basis of recent evidence, including his original studies, Susuki indicates that dysfunction or disruption of nodes of Ranvier play significant roles in the pathogenesis of immune-mediated neuropathies (Chap. 8). Kato et al. describes Schwann cell abnormalities in relationship to the underlying causes of diabetic neuropathy, such as polyol pathway hyperactivity, protein kinase C activity abnormalities, oxidative stress, glycation, and impaired neurotrophic support (Chap. 9).

1.4 Spontaneously Immortalized Schwann Cell Lines from Adult Rodents for the Study of Peripheral Nerve Degeneration and Regeneration (Chap. 10)

Watabe et al. have established spontaneously immortalized Schwann cells from adult normal mice (e.g., IMS32) (Watabe et al. 1995) and rats (e.g., IFRS1) (Sango et al. 2011), as well as murine disease models (Watabe et al. 2003). The methodology for establishing these cell lines and their characteristic features and usefulness for the study of nerve degeneration and regeneration are summarized in Chap. 10.

1.5 Conclusion

Thanks to all the contributors, this book presents wide aspects of Schwann cells from basic to clinical neuroscience, including their biological significance under normal and pathological conditions, and recent progress in the molecular techniques for approaching the events where Schwann cells have critical roles. We hope that this book will help the readers reconfirm the multifunctional roles of Schwann cells as glial cells in the peripheral nervous system.

References

Sango K, Yanagisawa H, Kawakami E, Takaku S, Ajiki K, Watabe K (2011) Spontaneously immortalized Schwann cells from adult Fischer rat as a valuable tool for exploring neuron–Schwann cell interactions. J Neurosci Res 89(6):898–908

Torii T, Miyamoto Y, Onami N, Tsumura H, Nemoto N, Kawahara K, Kato M, Kotera J, Nakamura K, Tanoue A, Yamauchi J (2013) In vivo expression of the Arf6 guanine-nucleotide exchange factor cytohesin-1 in mice exhibits enhanced myelin thickness in nerves. J Mol Neurosci 51(2):522–531

Watabe K, Fukuda T, Tanaka J, Honda H, Toyohara K, Sakai O (1995) Spontaneously immortalized adult mouse Schwann cells secrete autocrine and paracrine growth-promoting activities. J Neurosci Res 41(2):279–290

Watabe K, Sakamoto T, Kawazoe Y, Michikawa M, Miyamoto K, Yamamura T, Saya H, Araki N (2003) Tissue culture methods to study neurological disorders: establishment of immortalized Schwann cells from murine disease models. Neuropathology 23(1):68–78

Yamauchi J, Miyamoto Y, Torii T, Takashima S, Kondo K, Kawahara K, Nemoto N, Chan JR, Tsujimoto G, Tanoue A (2012) Phosphorylation of cytohesin-1 by Fyn is required for initiation of myelination and the extent of myelination during development. Sci Signal 5(243):ra69

Chapter 2
Recent Insights into Molecular Mechanisms That Control Growth Factor Receptor-Mediated Schwann Cell Morphological Changes During Development

Yuki Miyamoto and Junji Yamauchi

Abstract Rapid progress in molecular biological techniques, especially for gene modification technologies as well as RNA technologies, allows us to investigate how Schwann cells differentiate and myelinate axons at in vitro and in vivo levels, at the same rate until an experimental conclusion is reached. Through a middle embryonic stage after birth, Schwann cell myelination processes are categorized into three stages: cell migration, cell elongation along axons, and axonal myelination. However, these stages are not independent of each other but are a series of continuous processes. Herein, we review recent progress in unique, intracellular molecular mechanisms focusing on growth factor receptor-mediated Schwann cell morphological changes throughout the myelination processes, possibly identifying one basic mechanism of forming Schwann cell morphology. Such studies also provide a basis for understanding how human pathological states arise.

Keywords Differentiation • Gene modification technology • Migration • Molecular mechanism • Myelination • Schwann cell • Signaling • Small GTPase kinase

2.1 Introduction

Neural crest cells in the dorsal neural tube generate Schwann cell precursors, which in turn give rise to Schwann cells. During development of the peripheral nervous system, Schwann cell lineage cells migrate along axons to their final destination, where mature Schwann cells eventually wrap around individual axons to form a myelin sheath. The myelin sheath is the morphologically differentiated Schwann cell plasma membrane that insulates axons and markedly increases nerve

Y. Miyamoto • J. Yamauchi (✉)
Department of Pharmacology, National Research Institute for Child Health and Development, 2-10-1 Okura, Setagaya, Tokyo 157-8535, Japan
e-mail: yamauchi-j@ncchd.go.jp

K. Sango and J. Yamauchi (eds.), *Schwann Cell Development and Pathology*,
DOI 10.1007/978-4-431-54764-8_2, © Springer Japan 2014

conduction velocity (see Chaps. 3 and 4). The myelin sheath also plays a key role in protecting axons from various physical stresses, infections, and excessive immune reactions (see Chaps. 5 and 8).

In addition to the myelinating Schwann cell, Schwann cells are further categorized into three groups: the nonmyelinating Schwann cell, the perisynaptic Schwann cell, and the satellite cell. These cells are all derived from neural crest cells. Although myelination of axons by the Schwann cell is an event that only occurs in large-diameter axons (more than 1 μm in diameter), the nonmyelinating Schwann cell can surround the small-diameter axons. The majority of the surrounding axons are composed of C-fiber axons, which originate from peripheral ganglia and consist of sympathetic and sensory neurons (see Chaps. 4, 5, and 10). The perisynaptic Schwann cell structurally and functionally helps to bridge the nerve terminal with its peripheral tissue to form the triparticle structures such as neuromuscular junctions. The satellite cell primarily associates with cell bodies positioned in peripheral ganglia and plays a role in separating the respective cell body units. There are several reviews on the categories of Schwann cells and each developmental process of the Schwann cell linage cell (Bunge 1993; Jessen and Mirsky 2005; Nave and Salzer 2006; Taveggia et al. 2010; Raphael and Talbot 2011; Pereira et al. 2012).

The term mature Schwann cell is unlikely to be quite different from the term Schwann cell in describing their involvement in the myelination process. Thus, we will simply use "Schwann cell" in later sentences.

Through a middle embryonic stage after birth, Schwann cell myelination processes are categorized into three phases: (1) cell migration, (2) cell elongation along axons, and (3) myelination. These phases are not independent of each other but are a series of continuous processes, for example, phases (1) and (2) are often called the premyelinating process (Figs. 2.1, and 2.2). Many studies allow us to understand "input signals" and "output signals" (for example, membrane receptors and nuclear factors, and vice versa) in the myelination processes. Many studies have shown that the myelination processes are linked to specific transcription factors and protein factors controlling general and specific chromatin remodeling, as well as proteins involved in various levels of RNA processing, such as messenger RNAs (mRNAs) and noncoding RNAs (ncRNAs) (Nave and Salzer 2006; Taveggia et al. 2010; Raphael and Talbot 2011; Pereira et al. 2012). However, comparatively less is known about intracellular proteins linked to cell-surface receptors and their regulation in the signaling pathway that couples input signals to output signals, especially for growth factor receptor signaling. Also, recent evidence suggests that molecular mechanisms controlling morphological changes and pathological mechanisms are closely tied to each other (see Chaps. 5, 6, 7, 8, and 9). Herein, we review recent progress in unique, intracellular molecular mechanisms, focusing on growth factor receptor-mediated Schwann cell morphological changes throughout the myelination processes, and provide a basis for understanding how human pathological states arise.

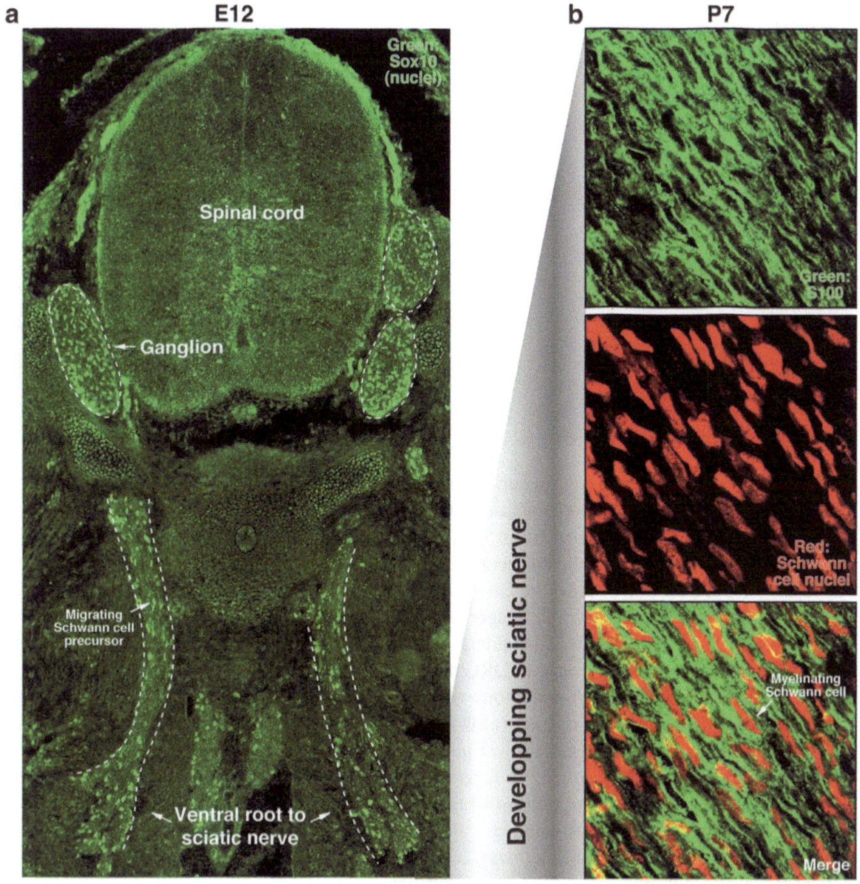

Fig. 2.1 Schwann cell lineage cells in mouse embryonic ganglia and neonatal sciatic nerves in vivo. (**a**) The spinal cord and ventral root linked to the sciatic nerve were exposed in a mouse embryo (embryonic day 12, E12). A thin section was fixed, blocked, and immunostained with an anti-Sox10 antibody, a nuclear marker for Schwann cell lineage cells (*green*). Sox10-positive Schwann cell precursors migrate from ganglia to peripheral tissues along ventral roots. (**b**) The mouse sciatic nerve (postnatal day 7, P7) was longitudinally cut, fixed, blocked, and co-stained with an S-100beta antibody (Schwann cell marker, *green*) and EthD-2 dye (nuclear staining, *red*). After birth, Schwann cells actively begin to wrap around axons with myelin sheaths. Production of myelin in mice is active until approximately 2 months

2.2 Molecular Processes of Migration, the Hallmark of Cell Morphological Changes, in Schwann Cell Lineage Cells

In middle to later embryonic stages in mouse or rat, a group of proliferating neural crest cells migrates from regions near the neural tubes to ganglia. Schwann cell precursors then proliferate as they migrate toward the periphery along growing axons

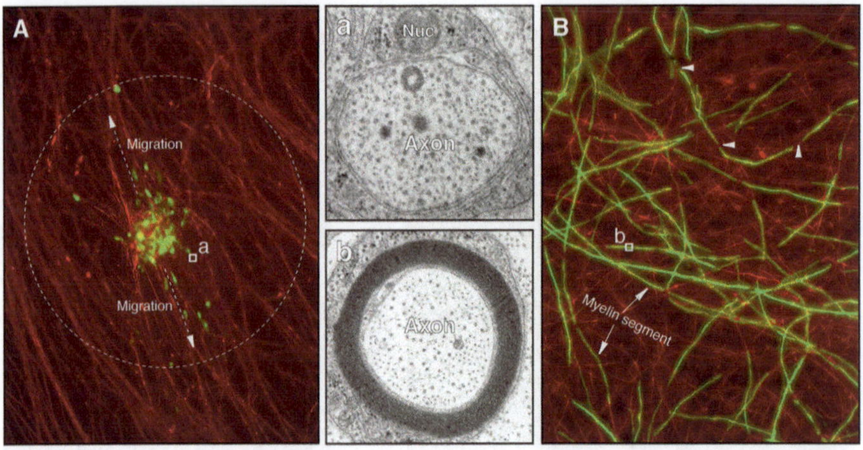

Fig. 2.2 Migrating Schwann cell precursors on axons and myelination of axons by Schwann cells in vitro. **A** Rat Schwann cell precursors and DRG neurons were purified from E14.5 dorsal root ganglions (DRGs). *Green fluorescent* protein-expressing, reaggregated Schwann cell precursors were put on DRG axons (*red* indicates neurofilament subunits) and allowed to migrate along axons outwardly (in the direction of the *dotted arrows*). *a* (*middle upper micrograph*) shows cross section (representative electron microscopic image; *nuc* indicates Schwann cell nucleus) of *small panel a* in *large panel A*. In *a*, Schwann cells undergo elongation on axons before initiating myelination. **B** Schwann cells were co-cultured with DRG neurons and allowed to undergo myelination. Myelin segments (indicated by *arrows*) and axons were immunostained with antibodies against myelin basic protein (MBP) (*green*) and neurofilaments (*red*). *Arrowheads* indicate myelin domains between myelin segments. *b* (*middle lower micrograph*) is a cross section (representative electron microscopic image) of *small panel b* in *large panel B*. In *b*, Schwann cells undergo myelination. Importantly, time-courses of migration (**A**) and myelination (**B**) in vitro are almost the same as those in vivo. Thus, experiments using co-culture techniques are now being applied to studies on analyzing the signal transduction pathway controlling migration and myelination. Experimental procedures for co-cultures are explained in Chap. 10

before initiating myelination after birth. However, migration is not fully understood. For example, it is not known whether proliferation and migration phenomena are happening at the same time or at independent stages, or if proliferation mostly precedes migration similar to precursors in other cell type lineages.

Genetic studies in zebrafish have shown how proliferation is associated with migration (Lyons et al. 2005; Pogoda et al. 2006; Raphael and Talbot 2011). Zebrafish were used as a model rather than rodents because zebrafish are visually transparent. Several years ago, the group of Talbot et al. carried out a screen for mutants with disrupted transcription of the *myelin basic protein* (*mbp*) gene, which is the major myelin protein gene (Raphael and Talbot 2011). Genetic mapping in mutants has first identified the zebrafish orthologues ErbB2 and ErbB3. Zebrafish ErbB2 mutants have also displayed a phenotype with disrupted axons. Heterodimeric ErbB2 and ErbB3 tyrosine kinase receptors are expressed in Schwann cells. It is well known that they are essential regulators governing many aspects of myelination processes in mice (Riethmacher et al. 1997; Britsch et al. 1998; Morris et al. 1999;

Woldeyesus et al. 1999; Garratt et al. 2000; Citri et al. 2003; Nave and Salzer 2006; Taveggia et al. 2010). Genetic studies with pharmacological inhibition of ErbB receptors have provided evidence that signaling through ErbB receptors is essential for proliferation of Schwann cell precursors in ganglia and that they are required for directed migration along axons. In addition, inhibition of ErbB receptors has not affected the motility of Schwann cell precursors (Lyons et al. 2005). It is also noteworthy that postmigratory cells around zebrafish axons or peripheral tissues undergo an additional, final round of cell division, which is required for initiating myelination. In general, these facts are also consistent with the findings from zebrafish mutants that harbor a missense mutation in the ErbB receptor ligand, neuregulin-1 type III (Perlin et al. 2011). This type III ligand is essential for Schwann cell migration and ensures that peripheral glial cells are present in the correct numbers and positions during development. At present, although it is unknown whether proliferation of postmigratory cells occurs in mammalian Schwann cell lineage cells, it is certain that genetic studies in zebrafish will potentially provide us with a new concept that proliferation after migration may be required to proceed the myelination program.

Cyclin-dependent kinases and their kinase regulators are essential for regulating cell-cycle progression. Among them, the prototypic kinase Cdc2 phosphorylates vimentin in rodent Schwann cells (Chang et al. 2012). Vimentin is the main intermediate filament in Schwann cells (Chang et al. 2012; Triolo et al. 2012). The phosphorylation reaction accompanies phosphorylation of focal adhesion kinase (FAK) with upregulation of the activities of extracellular signal-regulated kinases (ERK) 1 and 2. The kinase reaction series results in activating the Schwann cell inside-out signaling of beta-1 integrins, resulting in promoting migration. Because active Cdc2 kinase is a general inducer to trigger cell-cycle progression, it is thought that proliferation likely occurs simultaneously with migration in mammalian Schwann cells. If this is correct, Schwann cell development in mammals may somewhat differ from that of zebrafish, in which proliferation likely occurs independently of migration. Further studies will increase our understanding of the precise relationship between proliferation and migration and of how proliferation may be responsible for migration and vice versa.

Primary Schwann cell precursors are suitable for studying cell migration and the related morphological changes in mammals; however, primary Schwann cells are often used in these studies, because their isolation is comparatively easier than that of Schwann cell precursors. In this section, we discuss both findings from Schwann cell precursors and Schwann cells as the Schwann cell lineage cell model.

2.2.1 Neuregulin-1 (NRG1) Promotes Migration

Neuregulin-1 (NRG1), which is provided by neurons such as ganglion neurons, is a potent mitogen for Schwann cells and promptly promotes migration (Bunge 1993; Jessen and Mirsky 2005; Nave and Salzer 2006; Taveggia et al. 2010; Raphael and

Talbot 2011; Pereira et al. 2012). NRG1 effects are enhanced by synergistic action with an intracellular increase of cyclic AMP (Raphael and Talbot 2011; Pereira et al. 2012). Schwann cells express proto-oncogene products ErbB2 and ErbB3 as the NRG1 receptor (Garratt et al. 2000; Citri et al. 2003; Lemke 2006; Macklin 2010). The neuronal membrane-bound NRG1 type III ligand (type III is probably the major NRG1 alternative splicing form in promoting migration in vivo) is activated through a complex process with extracellular proteases such as beta-site amyloid precursor protein-cleaving enzyme 1 (BACE1) and likely tumor necrosis factor-alpha converting enzyme (TACE) (Fleck et al. 2013). Mature NRG1 binds to ErbB3 and triggers the intrinsic ErbB2 tyrosine kinase activation and the downstream signaling. Although ErbB3 has been thought to be a kinase activity-deficient receptor, ErbB3 itself is demonstrated to have a low activity to cause autophosphorylation required to form various signaling complexes with the receptor in the intracellular domains (Shi et al. 2010).

Mahanthappa et al. (1996) originally identified NRG1 as the major ligand that stimulates Schwann cell migration in culture. Following stimulation with NRG1, activated ErbB2 binds, tyrosine phosphorylates, and activates the guanine-nucleotide exchange factor (GEF) Dock7, which results in increasing Schwann cell migration on axons (Fig. 2.3) (Yamauchi et al. 2008a, b). Dock7 is the GEF for Rac1 or Cdc42. Rac1 and Cdc42 belong to small GTPases of the Rho family and directly regulate actin and tubulin cytoskeletal proteins through various effector molecules. Similar to Ras GTPases, Rho GTPases act as molecular switches: they are biologically active when bound to GTP and are inactive when bound to GDP (Schmidt and Hall 2002; Rossman et al. 2005). Two types of specific proteins, GEFs and GTPase-activating proteins (GAPs), regulate their guanine nucleotide-binding states. The former reaction is important, because GEFs define the specificity and strength of GTPase activation by integrating upstream signals. GEFs for Rho GTPases are composed of approximately 80 Dbl family GEFs and 11 Dock180-related proteins (Schmidt and Hall 2002; Rossman et al. 2005; Matsuda and Kurata 1996; Miyamoto and Yamauchi 2010). Dock7 is a later subfamily member that is abundantly expressed in Schwann cells.

In migrating Schwann cells, active GTP-bound Rac1 and Cdc42, which act through their effector c-Jun N-terminal kinase (JNK) kinase kinase (JNKKK), activate JNK, which leads to phosphorylation of a focal adhesion protein, paxillin, as the JNK substrate (Miyamoto et al. 2012). Phosphorylation of paxillin by JNK is thought to stimulate turnover of focal adhesions to promote migration (Turner 2000; Huang et al. 2004; Yamauchi et al. 2006, 2007, 2008a, b).

When Schwann cells stop migration and are allowed to be placed along axons and elongated, the low-affinity nerve growth factor receptor $p75^{NTR}$ is aligned on the front of the premyelinating Schwann cell plasma membrane. It is known that molecules that are aligned with $p75^{NTR}$ in the front are the polarity proteins Par3/Par6/aPKC and they are directly regulated by Rac1 and Cdc42 (Chan et al. 2006; Chan 2007; Tep et al. 2012). The $p75^{NTR}$ contributes to wrapping axons in the myelin sheaths (see Sect. 2.3.3 for details). After birth, both the expression level and activity of JNK are greatly decreased (Parkinson et al. 2004); when Schwann cells are

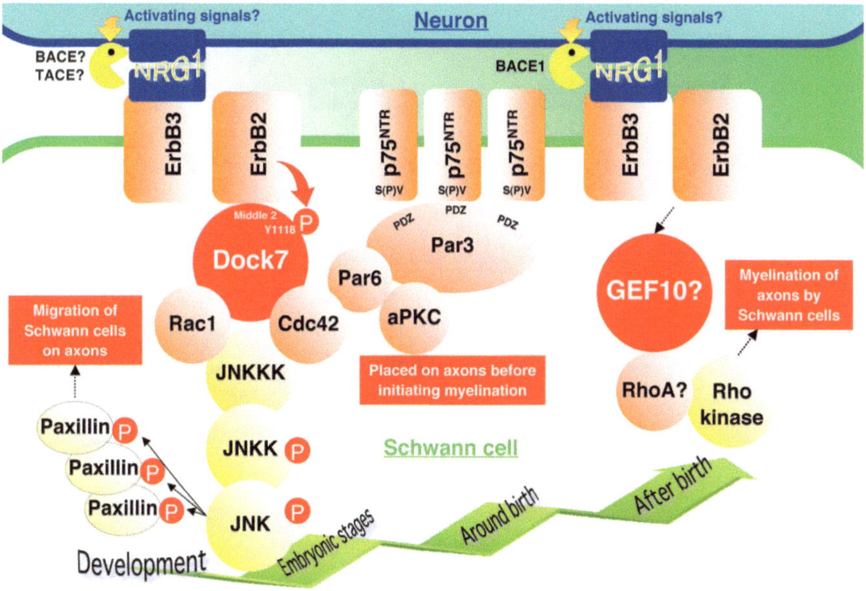

Fig. 2.3 Regulation of migration and early myelination phases by Dock7, the guanine-nucleotide exchange factor (GEF) for Rho GTPases Rac1 and Cdc42, in the Schwann cell–neuronal interaction. Rho GTPases Rac1 and Cdc42 are utilized as molecules in the signaling mechanisms by which rodent Schwann cell precursors migrate. Processed NRG1 binds to ErbB3 and leads to ErbB2 activation. ErbB2 directly binds Dock7, a GEF for Rho GTPases Rac1 and Cdc42, and phosphorylates it at the Tyr-1118 position. Phosphorylation stimulates the GEF activity of Dock7. Rac1 and Cdc42 directly activate their effector, JNK upstream kinase(s), and lead to phosphorylation of focal adhesion protein paxillin at the Ser-178 position through JNK. Paxillin phosphorylation is necessary for the recycling of focal adhesion proteins to promote migration. Around birth, JNK expression levels and activities are downregulated and, instead, Rac1 and Cdc42 localize $p75^{NTR}$-interecting polarity protein complex to the front of the Schwann cell plasma membrane. Thus, $p75^{NTR}$ can bind neuron-secreting brain-derived neurotrophin (BDNF), which is required for myelination. Afterward, expression levels and activities of Dock7 are downregulated. Rac1 and Cdc42 keep their activities through unidentified GEF(s) at the middle levels. In contrast, RhoA and the effector Rho-kinase activities are greatly upregulated during myelination. The Charcot-Marie-Tooth disease-responsible gene product, GEF10, is a candidate for RhoA-GEF (see Chap. 4). Further studies will clarify the complete picture for signaling through Rho GTPases

placed on axons, effectors for Rac1 and Cdc42 change from the JNK upstream kinase to the polarity protein complex. Conversely, effector exchange may cause Schwann cell elongation along axons before initiating myelination.

The nonreceptor tyrosine phosphatase Shp2 is known to be involved in NRG1 signaling, possibly through direct binding to ErbB2 and ErbB3 (Grossmann et al. 2009). Shp2 is a unique phosphatase that acts as an activator of the Src family nonreceptor tyrosine kinase (Hunter 2009). Shp2 can dephosphorylate the Tyr residue corresponding to Tyr-527 in c-Src at the C-terminus. Phosphorylation of this position is a negative regulation mechanism that is common in Src family kinases. Fyn is the major Src family kinase expressed in Schwann cells (Hossain et al. 2010)

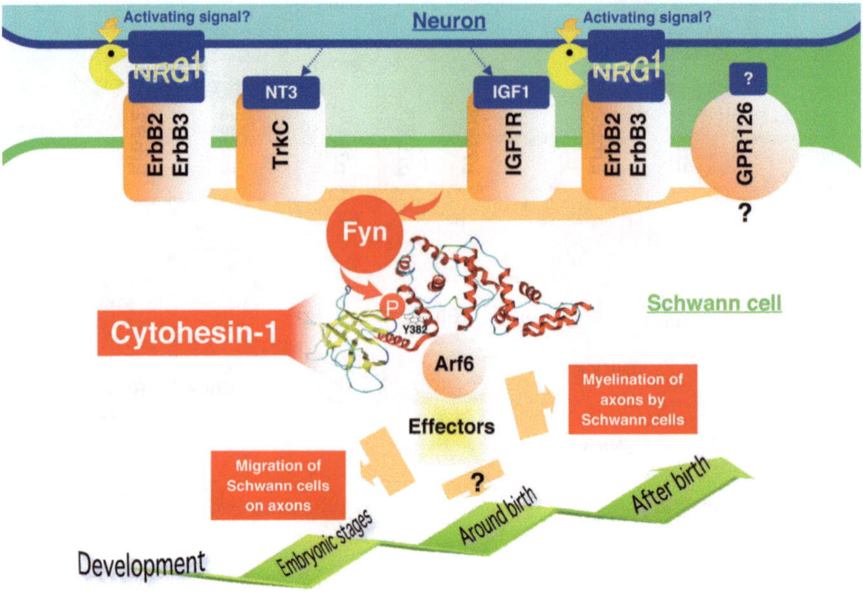

Fig. 2.4 Cytohesin 1, the GEF for Arf6, regulates myelination processes in the Schwann cell–neuronal interaction. Arf6, a small GTPase branch member, is utilized as one of the molecules in the mechanisms by which rodent Schwann cell precursors migrate and also of those in the mechanisms underlying myelination. In both cases, cytohesin-1 is the primary GEF for Arf6 in Schwann cells. Despite the specific involvement of Dock7 in ErbB2 and ErbB3 receptor signaling, cytohesin-1 is activated following stimulation of either of Schwann cell attractive receptors ErbB2/3, TrkC, or IGF1R during migration and following stimulation of either ErbB2/3 or IGF1R during myelination. The reason is probably that the Src family nonreceptor tyrosine kinase Fyn acts as a common signal mediator downstream of these receptors. It will be interesting to determine whether GPR126 also acts upstream of Fyn. Activated Fyn phosphorylates cytohesin 1 at the Tyr-382 position, resulting in activation of its GEF activity. Because Arf6 is required for almost every basic aspect of cell morphological changes in mammalian cells, Arf6 should contribute to both migration and myelination; however, it remains unclear which Arf6 effector is involved in migration or myelination

and has a critical role in migration at embryonic stages in the mouse (Yamauchi et al. 2012; Miyamoto et al. 2013). Thus, Shp2 may act upstream of Fyn in Schwann cells. Interestingly, it is known that NRG1-activated Fyn kinase phosphorylates and activates cytohesin-1 (Liu and Pohajdak 1992; Meacci et al. 1997; Kolanus et al. 1996; Geiger et al. 2000), which is the GEF for Arf6 in Schwann cells (Fig. 2.4). This phosphorylation reaction is the first example of the Arf family GEF as a functional tyrosine kinase substrate. Arf6 governs the cytoskeletal rearrangement and vesicular trafficking. Similar to Rho GTPases, Arfs, which are specifically controlled by their respective GEFs and GAPs, also act as molecular switches depending on their guanine nucleotide-binding states (D'Souza-Schorey and Chavrier 2006; Kahn et al. 2006; Casanova 2007; Donaldson and Jackson 2011).

Although it is not yet fully understood how NRG1 activation of Fyn promotes migration through cytohesin-1 and Arf6, migration generally requires cytoskeletal rearrangement and vesicular trafficking.

2.2.2 Antagonistic Regulation of Migration by Neurotrophins

Sensory neurons in dorsal root ganglion (DRG) neurons secrete neurotrophin-3 (NT3) and brain-derived neurotrophin (BDNF), which control Schwann cell migration on their axons. When Schwann cells are co-cultured with DRG neurons, the neurons secrete high levels of NT3 (Chan et al. 2001). In this initial state, Schwann cells are highly motile. NT3 binds to the cognate receptor TrkC on Schwann cells and promotes migration through Rac1 and Cdc42 and the downstream activation of JNK (Yamauchi et al. 2003; Miyamoto et al. 2012). In this case, Tiam1 (Dbl family GEF) specifically activates Rac1, whereas Dbs (Dbl family GEF) specifically activates Cdc42 (Yamauchi et al. 2005a, b; Miyamoto et al. 2006). NT3-activated TrkC can come to bind, tyrosine phosphorylate, and activate Cdc42-GEF Dbs, which is the first example of the GEF as a tyrosine kinase receptor-binding partner (Yamauchi et al. 2005a). It is important to note each GEF activates each GTPase in neurotrophin signaling, differing from NRG1, which activates both Rac1 and Cdc42 through Dock7.

Activated Rac1 and Cdc42 are also involved in formation of the polarity protein complex containing Par3/Par6/aPKC (Chan et al. 2006). It is noteworthy that the mechanisms downstream of Rac1 and Cdc42 are conserved in premyelinating phases (Miyamoto et al. 2012), as is also seen with cytohesin-1 and Arf6 (Fig. 2.4) (Miyamoto et al. 2013).

Conversely, BDNF expression levels in a culture medium are constant during Schwann cell–DRG neuronal co-cultures. As myelination begins, NT3 is gradually downregulated and BDNF becomes the major neurotrophin during myelination. These phenomena occur in mice (Cosgaya et al. 2002). Schwann cells do not express the high-affinity BDNF receptor TrkB; instead, BDNF binds to $p75^{NTR}$ (low-affinity pan-neurotrophin receptor) and inhibits migration before initiating myelination. The repulsive effect through BDNF binding to $p75^{NTR}$ is intracellularly mediated by RhoA activation through RhoA-GEF Vav2 (Yamauchi et al. 2004), which signals through the interaction with the $p75^{NTR}$-associating protein, tumor necrosis factor receptor-associated factor 6 (TRAF6), and the Src family in Schwann cells (Cosgaya and Shooter 2001; Yamauchi et al. 2004).

2.2.3 Insulin-Like Growth Factor 1 (IGF-1) Promotes Migration

Although insulin-like growth factor 1 (IGF-1) is known to be a key regulator of myelination, its role in migration is also important (Ogata et al. 2004, 2006).

In culture, IGF-1 activates the IGF-1 receptor (IGF-1R) on Schwann cells and couples insulin receptor substrate 1 (IRS-1) and phosphatidylinositol-3-kinase (PI3K) to Rho GTPases Rac1 and Cdc42 and JNK, which promotes migration (Cheng et al. 2000; Miyamoto et al. 2012). IGF-1 signaling also involves cytohesin-1 and Arf6 (Fig. 2.4) (Miyamoto et al. 2013), suggesting that Schwann cells use a common molecular mechanism for migration.

2.2.4 Glial Cell Line-Derived Neurotrophic Factor (GDNF) Promotes Migration

The functional receptor for glial cell line-derived neurotrophic factor (GDNF) is neural cell adhesion molecule (NCAM) in Schwann cells (Paratcha et al. 2003; Zhou et al. 2003; Iwase et al. 2005). When NCAM interacts with the GDNF family receptor (GFR) alpha-1, the ability of NCAM to promote cell adhesion is decreased, whereas GDNF binding to NCAM is enhanced; their events activate Fyn and FAK to promote migration. Importantly, analyses using RET−/− mice demonstrate that the known GDNF receptor, RET, is not required for Schwann cell migration.

However, it is unlikely that GDNF is the primary regulator for migration in an in vitro explant assay system (Heermann et al. 2012). Conversely, the fact that GDNF primarily regulates migration may result from the potency of NRG1 and neurotrophins, which are physiological factors for migration during development.

2.2.5 Growth Arrest-Specific 6 (Gas6) Promotes Migration

Growth arrest-specific 6 (Gas6) is a ligand for TAM tyrosine kinase receptors (Tyro3, Axl, and Mer receptors). Among these receptors, Tyro3 is highly expressed in human Schwann cells whereas Gas6 is expressed in human DRG neurons (Li et al. 1996). Indeed, Gas6 triggers Tyro3 receptor phosphorylation with ERK2 upregulation in Schwann cells. However, together with the potent Schwann cell mitogen NRG1, Gas6 supports maximal Schwann cell proliferation and migration. Gas6 itself can also preserve the premyelinating Schwann cell morphology, along with expression of the premyelinating Schwann cell markers such as S100-beta and glial fibrillary acidic protein (GFAP). Signaling through Gas6 and Tyro3 only has a supportive effect on morphological changes in premyelinating Schwann cells. Interestingly, Gas6 and TAM receptors were recently found to be responsible for autophagy (Nguyen et al. 2013). Thus, Gas6 and Tyro3 may constitute part of the molecular machines to undergo autophagy in Schwann cells (Rangaraju et al. 2010). The basic biological facts present a common mechanism between migration and autophagy (Reddien and Horvitz 2004).

2.2.6 Promotion of Migration Through the Low Density Lipoprotein (LDL) Receptor-Related Protein-1 (LRP1)

Some Schwann cell receptors are upregulated as development proceeds and also after injury. After peripheral nerve injury in mice, Schwann cells upregulate an endocytic receptor molecule called low density lipoprotein (LDL) receptor-related protein-1 (LRP1) (Mantuano et al. 2010). LRP1 is likely one of the mechanisms that allow Schwann cells to adopt a migratory phenotype to recover from injury. Knockdown of LRP1 in Schwann cells decreases migration, with upregulation of RhoA activity and downregulation of Rac1 activity. The Rho family molecular mechanism where RhoA inhibits migration and Rac1 promotes migration is conserved in Schwann cells (Yamauchi et al. 2004). Also, stimulation with either of an LRP1 ligand, such as matrix metalloprotease-9 (MMP9), or tissue-type plasminogen activator (tPA), or alpha-2-macroglobulin (APOE), promotes migration; however, it is not known that these ligands have their effects in an autocrine or paracrine manner in vivo. Signaling through LRP1 is also involved in Schwann cell morphological changes under multiple aspects of developmental and pathological states (Mantuano et al. 2011; Orita et al. 2013).

2.2.7 Promotion of Migration Through the Erythropoietin Receptor

Schwann cells upregulate both erythropoietin and erythropoietin receptor following nerve injury (Inoue et al. 2010). Erythropoietin not only stimulates migration in a manner dependent on Janus kinase 2 (JAK2), which is known to be the major erythropoietin receptor-associating protein, but also participates both in recruitment of migration-supporting beta-1 integrins to the cell surface and in fibronectin expression, illustrating that erythropoietin is a multifunctional growth factor after injury.

2.2.8 Regulation of Migration Through Semaphorin Signaling

It is well characterized that neuropilins and group A plexins are composed of receptor complexes for class 3 semaphorins and help to guide neural progenitor cell migration and growth cones in axons during central nervous system development. Interestingly, Schwann cells also contain mRNAs encoding semaphorin-3F (Sema3F) receptors neuropilin-2 (Nrp2) and plexin-A3 (PlexA3) (Ara et al. 2005). Nrp1, which also participates in semaphorin-3A (Sema3A) signaling, is expressed in Schwann cells. In Nrp-1−/− mutants, the number of Schwann cells migrating along axons is greatly reduced (Huettl and Huber 2011). It will be important to determine whether the repulsive or attractive effect of semaphorin signaling on Schwann cell migration displays the same responsiveness as that of semaphorin signaling on neuronal linage cell behaviors.

2.2.9 *Regulation of Migration Through Ephrin Signaling*

Besides a number of studies on the effects of Schwann cell grafts on the repulsive reaction of spinal cord astrocytes in a mouse model, Schwann cell–astrocyte interaction participates in establishing the boundary between peripheral nervous system and central nervous system. Afshari et al. (2010) reported that astrocytes express ephrinA1, ephrinA2, ephrinA3, ephrinA4, and ephrinA5, whereas Schwann cells express EphA2, EphA4, EphA7, EphB2, EphB3, and EphB6. Among them, the most effective combination is ephrinA5, as the ligand on astrocytes, and EphA4, as the receptor on Schwann cells. EphrinA5 binding to EphA4 upregulates the intrinsic EphA4 tyrosine kinase activity to activate Vav2 and RhoA, causing a repulsive response of Schwann cells against astrocytes. The mechanism of this repulsive response through EphA4 is the same as that for $p75^{NTR}$ signaling (Yamauchi et al. 2004). Vav2 and RhoA may form one of the common signaling units responding to repulsive cues in Schwann cells.

2.3 Molecular Processes of Myelination of Axons by Schwann Cells

Axonal myelination by Schwann cells begins after birth. Over time, myelin sheaths can grow to be more than one hundred times larger than the collective surface area of the premyelinating Schwann cell plasma membrane. This process is one of the most dynamic morphological differentiations in mammalian cells (Bunge 1993; Jessen and Mirsky 2005; Nave and Salzer 2006; Taveggia et al. 2010; Raphael and Talbot 2011; Pereira et al. 2012). Before discussing molecular mechanisms underlying growth factor receptor-mediated myelination, we discuss the relationship between migration and myelination or their transition.

Although Cdc2 kinase and the substrate vimentin promote migration, vimentin knockout mice display a hypermyelination phenotype in the peripheral nervous system (Triolo et al. 2012), indicating that vimentin negatively regulates myelination. Signaling through Cdc2 may help Schwann cells to sustain a promigratory state. Downregulation may provide Schwann cells with myelination properties. The cyclin-dependent kinase inhibitor p57kip2 is also involved in progression to the myelination phase. In general, it plays a role in blocking the G_1–S transition throughout the cell cycle. Knockdown of p57kip2 in culture leads to cell-cycle exit, actin filament stabilization, and altered cell morphology, being associated with progression of the myelination program (Heinen et al. 2008). It is suggested that fine-tuned regulation among cell-cycle inhibitors triggers the switch to myelination. More broadly, downregulation of potent, proliferative signal(s) or modulation of the inhibitor(s) or both may promote Schwann cell differentiation and myelination. The mechanisms controlling their expression involve growth factors and their receptor signaling. Here, we discuss axonal myelination in comparison with migration or as an extension line of mechanisms underlying migration.

2.3.1 NRG1 Promotes Myelination

After birth and the myelination phase, neurons are the only source of NRG1 (Bunge 1993; Jessen and Mirsky 2005; Nave and Salzer 2006; Taveggia et al. 2010; Raphael and Talbot 2011; Pereira et al. 2012), and Schwann cells express ErbB2 and ErbB3 as the NRG1 receptor (Garratt et al. 2000; Citri et al. 2003; Lemke 2006; Macklin 2010). Type III is the only major NRG1 alternative splicing form to promote myelination. In the myelination phase, NRG1 type III is specifically activated by membrane-bound protease BACE1 (Willem et al. 2006; Hu et al. 2008) and specifically inactivated by TACE (La Marca et al. 2011).

Mature NRG1 binds to ErbB3 and triggers the intrinsic ErbB2 tyrosine kinase activation. Dock7 is the key ErbB2 substrate while Schwann cells migrate, but it is not known whether its mechanism is preserved during myelination. When Schwann cells begin to myelinate, Dock7 expression and its phosphorylation state are promptly decreased. In vivo knockdown of Dock7 results in enhanced myelin thickness in sciatic nerves (Yamauchi et al. 2011). This finding illustrates that Dock7 is a negative regulator of myelination (Fig. 2.3). Dock7 potentially inhibits myelination through the upstream kinase(s) of JNK and p38 MAPK, because JNK and p38 mitogen-activated protein kinase (MAPK) negatively regulate myelination (Parkinson et al. 2004; Yamauchi et al. 2008a, b, 2011; Yang et al. 2012). The upstream kinase(s) activating JNK and p38 MAPK are the effectors of Rho GTPases Rac1 and Cdc42. It is thus likely that the activities of JNK and p38 MAPK are downregulated as myelination proceeds (Parkinson et al. 2004; Yang et al. 2012).

Although Dock7 is the GEF for Rac1 and Cdc42 in Schwann cells, Rac1 and Cdc42 are required not only for polarity protein complex formation before initiating myelination (Chan et al. 2006) but also for proper morphological changes of Schwann cells during myelination and Schwann cell axonal sorting (Benninger et al. 2007; Nodari et al. 2007). Rac1 activates the effector p21-activated kinases (PAKs). Activated PAKs can phosphorylate and activate Lin11, Isl-1, and Mec-3 domain kinases (LIMKs). LIMKs specifically phosphorylate cofilin, resulting in reorganization of the actin cytoskeleton that is required for myelination (Sparrow et al. 2012). Neuronal Wiskott–Aldrich syndrome protein (N-WASP) is a specific Cdc42 effector and reorganizes the actin cytoskeleton through the Arp2/3 complex. Schwann cell-specific N-WASP knockout mice display severe hypomyelination, indicating that N-WASP plays a key role in myelination (Jin et al. 2011; Novak et al. 2011). RhoA, however, is not activated during myelination (Yamauchi et al. 2011) and promotes myelination through the RhoA effector Rho-kinase (Melendez-Vasquez et al. 2004). Further studies in this line will enable us to identify ErbB2-responsible GEFs for Rho GTPases during myelination.

Fyn-induced phosphorylation of Arf6-GEF cytohesin-1 is also conserved in the myelination phase (Yamauchi et al. 2012). Similarly to the migration phase, ErbB2 and Shp2 are thought to act as Fyn membrane transduction molecules. In addition, FAK is the well-known binding partner of Fyn in focal adhesion sites, and it is also linked to ErbB2/3 and α_6-β_1 integrin (Chen et al. 2000; Scherer 2001; Feltri

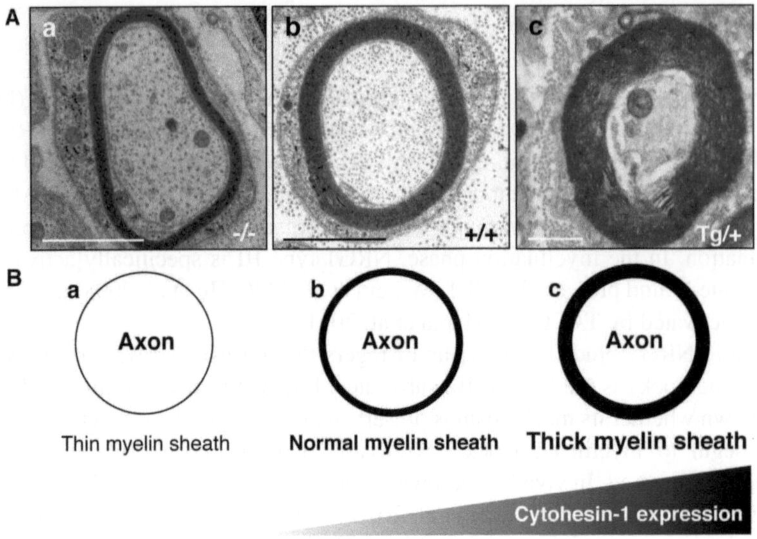

Fig. 2.5 Modulation of myelin thickness by artificially adjusting cytohesin 1 expression levels. (**A**) *a*, *b*, and *c* indicate the respective cross-section electron micrographs of postnatal day 7 sciatic nerves in cytohesin 1 knockout (–/–), wild type (+/+), and transgenic (Tg/+) mice. *Bars* 1 μm. (**B**) The cartoon models (*panels a*, *b*, and *c* correspond to those in **A**) illustrate that, as cytohesin 1 expression increases, the myelin sheath becomes thick, and myelin thickness depends on gene dosage (expression level) of cytohesin 1 (Yamauchi et al. 2012)

et al. 2001). FAK–/– mice display hypomyelination and defective axonal sorting by Schwann cells (Grove et al. 2007). Integrin-linked kinase (ILK) is also localized in focal adhesion sites and participates in myelination and axonal sorting. Thus, in Schwann cells, signalsome(s) including ILK, FAK, and Fyn, which act downstream of the membrane transduction molecules, may prepare to wrap around axons with myelin sheaths. Fyn is also activated following stimulation with other growth factors. Thus, cytohesin 1 and Arf6 are among the common intracellular molecules controlling myelination (Fig. 2.4).

Interestingly, as expression levels of cytohesin 1 are increased, the myelin sheaths formed become thicker (Fig. 2.5) (Yamauchi et al. 2012). This molecular property is similar to that observed with NRG1 type III (Nave and Salzer 2006).

2.3.2 *Control of Myelination Through a Possible Cyclic AMP-Producing G Protein-Coupled Receptor 126 (GPR126)*

Upregulation of intracellular cyclic AMP is essential for myelination. In the absence of NRG1, the cell-permeable cyclic AMP analogue can mimic progression of the myelination program in experiments using primary Schwann cells (Jessen and Mirsky 2005; D'Antonio et al. 2006). Genetic studies using zebrafish have identified a possible cyclic AMP-producing G protein (Gs)-coupled receptor 126 (GPR126) (Monk et al. 2009). In GPR126 mutants, zebrafish Schwann cells do not express the myelin protein expression-associated transcription factors Oct6 and Krox20, and phenotypes are arrested at the promyelinating stage. These mutants were partially reversed by addition of cyclic AMP. Thus, zebrafish GPR126 is responsible for driving differentiation of promyelinating Schwann cells. Similar results are observed in GPR126+/− mice, which display decreased GPR126 expression (Monk et al. 2011). Because GPR126 complete knockout mice are embryonic lethal, analyses using conditional knockout mice will be required to validate the precise function of GPR126 in myelination. In addition, it is also not known whether cyclic AMP activates protein kinase A (PKA) (a classical cyclic AMP target) or GEFs for Rap GTPases (new targets) to enhance myelination (Gloerich and Bos 2010).

2.3.3 Antagonistic Regulation of Myelination by Neurotrophins

When Schwann cells are co-cultured with DRG neurons, the neurons secrete NT3. As myelination begins, NT3 is gradually downregulated. Because BDNF secretion levels from neurons are constant during co-culture, BDNF becomes the main neurotrophin as myelination proceeds. Similar changes of neurotrophins occur in mice (Chan et al. 2001; Cosgaya et al. 2002). As a result, NT3 activation of TrkC inhibits myelination whereas BDNF binding to $p75^{NTR}$ promotes myelination. The reciprocal effect of these neurotrophins for myelination is seen for migration (see Sect. 2.2.2). In addition, nerve growth factor (NGF) may also be an endogenous ligand of $p75^{NTR}$ (Chan et al. 2004). In either case, Schwann cell $p75^{NTR}$ is one of the main regulators of myelination.

2.3.4 Promotion of Myelination by IGF-1 and Its Inhibition by Platelet-Derived Growth Factor (PDGF)

IGF-1 is the key positive regulator of Schwann cell myelination, as well as migration (Ogata et al. 2004, 2006). In experiments using co-cultures and allogeneic nerve graft experiments, IGF-1 promotes myelination through the glycogen synthase kinase 3 (GSK3) beta-Akt cascade. IGF-1 also promotes it through the pathway coupling Arf6 to PI3K and Akt (Fig. 2.4) (Donaldson and Jackson 2011; Yamauchi et al. 2012). In either case, Akt has a key role in myelination.

Platelet-derived growth factor (PDGF) is antagonistic to the action of IGF-1 through a robust upregulation of ERK (Ogata et al. 2004; Monje et al. 2009). Apra et al. (2013) found that collagen triple helix repeat containing 1 (Cthrc1) is a potent mitogenic factor (and possibly a migrating factor) for Schwann cells and that it negatively regulates myelination. Cthrc1 may be a secretory protein, although it contains typical ERK phosphorylation sites (see the Scansite website: http://scansite.mit.edu/). Cthrc1 may act as the ERK substrate downstream of the PDGF receptor. However, in some cases, ERK is proposed to be the positive regulator of myelination. The effect of ERK on myelination likely differs in a context-dependent manner in Schwann cells in vitro and in vivo (Syed et al. 2010; Newbern et al. 2011; Napoli et al. 2012; Ishii et al. 2013).

In addition, Akt activation is antagonistically regulated by the myotubularin-related proteins Mtmr2 and Mtmr13 in Schwann cells (Ng et al. 2013). Mtmr2 and Mtmr13 are gene products of autosomal recessive demyelinating peripheral neuropathy Charcot-Marie-Tooth disease type 4B1 and 4B2 (CMT4B1 and CMT4B2; Bolino et al. 2000), which are characterized by myelin outfolding and axonal degeneration (see Chap. 4). Mtmr2 possesses specific phosphatase activity toward PI3K products phosphatidylinositol-3-phosphate and phosphatidylinositol-3,5-bisphosphate. Thus, Mtmr2 blocks Akt activation by dephosphorylating PI3K products. Although Mtmr13 is a pseudophosphatase, Mtmr2 and Mtmr13 depend on each other to maintain their protein expression levels, possibly by assisting with their protein stability. Mtmr2 and Mtmr13 may act downstream of the PDGF receptor in Schwann cells.

Signaling through cytohesin 1, which also participates in myelination regulation by IGF-1 and neurotrophins, is summarized in Fig. 2.4.

2.3.5 GDNF Promotes Myelination

In co-cultures, GDNF enhances myelination in a manner dependent on the Schwann cell functional receptor NCAM, as observed during migration (Iwase et al. 2005). The effect is especially important in an early stage of myelin formation. It will be interesting to determine whether Fyn and FAK mediate NCAM-dependent myelination (Iwase et al. 2005) and also whether a mechanism underlying myelination is conserved in multiple receptors involving NCAM.

2.4 Conclusions and Perspectives

GEFs are emerging as important regulators throughout the myelination process. They define the specificity or the strength of GTPase activation and the following kinase reactions by integrating upstream signals. Therefore, they play key roles in many aspects of Schwann cell morphological changes.

To date, two GEFs for Rho GTPases are known to be associated with peripheral neuropathies (see Chap. 6). The Cdc42-specific GEF frabin [also known as FYVE, Rho-GEF and plecksrtrin homology domain containing 4 (FGD4)] is heterogeneously mutated in autosomal recessive Charcot-Marie-Tooth disease CMT4H.

All frabin mutations are presumed to disrupt protein conformational changes, which are required for Cdc42 activation or association with frabin-binding molecules such as actin and phospholipids, which cause aberrant myelin formation such as myelin outfolding and myelin dysfunction (Delague et al. 2007; Stendel et al. 2007). Another GEF responsible for peripheral neuropathy is RhoA-specific GEF Arfgef10 (also called GEF10). One missense mutation is associated with a peripheral neuropathy with slowed nerve conduction velocities, which has peripheral neuronal axons that are thinly myelinated by Schwann cells (Verhoeven et al. 2003). These mutations are probably associated with improper GEF regulation of Rho GTPases in Schwann cell morphogenesis during myelination. These findings expand our knowledge concerning novel molecular networks underlying the myelination process.

During the past two decades, many studies of Schwann cell development have clearly identified the required membrane transduction signals, such as the neuronal NRG1 ligand and the cognate receptors ErbB2 and ErbB3 on Schwann cells, and the signaling pathways that lead to the essential transcription factor cascade Sox10/Oct6/Krox20. Further studies in this area will create a new type of regenerative medicine using new biotechnological techniques. In oligodendrocytes, such studies are advancing at an increasing rate. Using transcription factor-mediated reprogramming, recent biotechnology methods enable embryonic rodent fibroblasts to directly convert induced oligodendrocyte progenitor cells, without passing through stem cell-like phenotypes (Najm et al. 2013; Yang et al. 2013). Induced oligodendrocyte progenitor cells have the potential to produce biologically active myelin sheaths. The next stage probably leads to direct in vivo reprogramming toward oligodendrocyte lineage cells (and also, of course, Schwann cell lineage cells), using the smallest set of transcription and signaling molecules or artificial bioorganic materials.

Increasing knowledge about these stem cell fields allows us to understand that *Mycobacterium leprae*, which causes leprosy, has the ability to reprogram adult Schwann cells to a stage of progenitor/stem-like cells to promote bacterial dissemination (Masaki et al. 2013). This finding may give us some hint for direct reprogramming to Schwann cell lineage cells.

In addition to development of the stem cell research, basic studies on glial–neuronal interactions have developed a glial biotechnology whereby engineered nanofibers can reproduce oligodendrocyte myelination without primary rodent neurons (Lee et al. 2012, 2013). Similar biotechnological methods will be applied to Schwann cell myelination in future studies. Further elucidation of basic mechanisms in Schwann cell development enables us to develop more practical regenerative medicine and also to reconsider the long-standing, attractive question of how Schwann cells achieve dynamic morphological changes and why Schwann cells have many abilities far beyond those of oligodendrocytes. Such studies will allow us to focus on developing therapeutic methods concerning myelin-related diseases and more general neuropathies.

In this chapter, we have focused on the current understanding of Schwann cell morphological changes and the related signaling mechanisms. We presented a common mechanism in each developmental stage, and showed there was some similarity between development and pathological progression (see Chaps. 5, 6, 7, 8, and 9).

References

Afshari FT, Kwok JC, Fawcett JW (2010) Astrocyte-produced ephrins inhibit Schwann cell migration via VAV2 signaling. J Neurosci 30:4246–4255

Apra C, Richard L, Coulpier F, Blugeon C, Gilardi-Hebenstreit P, Vallat JM, Lindner V, Charnay P, Decker L (2013) Cthrc1 is a negative regulator of myelination in Schwann cells. Glia 60:393–403

Ara J, Bannerman P, Shaheen F, Pleasure DE (2005) Schwann cell-autonomous role of neuropilin-2. J Neurosci Res 79:468–475

Benninger Y, Thurnherr T, Pereira JA, Krause S, Wu X, Chrostek-Grashoff A, Herzog D, Nave KA, Franklin RJ, Meijer D, Brakebusch C, Suter U, Relvas JB (2007) Essential and distinct roles for cdc42 and rac1 in the regulation of Schwann cell biology during peripheral nervous system development. J Cell Biol 177:1051–1061

Bolino A, Muglia M, Conforti FL, LeGuern E, Salih MA, Georgiou DM, Christodoulou K, Hausmanowa-Petrusewicz I, Mandich P, Schenone A, Gambardella A, Bono F, Quattrone A, Devoto M, Monaco AP (2000) Charcot-Marie-Tooth type 4B is caused by mutations in the gene encoding myotubularin-related protein-2. Nat Genet 25:17–19

Britsch S, Li L, Kirchhoff S, Theuring F, Brinkmann V, Birchmeier C, Riethmacher D (1998) The ErbB2 and ErbB3 receptors and their ligand, neuregulin-1, are essential for development of the sympathetic nervous system. Genes Dev 12:1825–1836

Bunge RP (1993) Expanding roles for the Schwann cell: ensheathment, myelination, trophism and regeneration. Curr Opin Neurobiol 3:805–809

Casanova JE (2007) Regulation of Arf activation: the Sec7 family of guanine nucleotide exchange factors. Traffic 8:1476–1485

Chan JR (2007) Myelination: all about Rac 'n' roll. J Cell Biol 177:953–955

Chan JR, Cosgaya JM, Wu YJ, Shooter EM (2001) Neurotrophins are key mediators of the myelination program in the peripheral nervous system. Proc Natl Acad Sci USA 98:14661–14668

Chan JR, Watkins TA, Cosgaya JM, Zhang C, Chen L, Reichardt LF, Shooter EM, Barres BA (2004) NGF controls axonal receptivity to myelination by Schwann cells or oligodendrocytes. Neuron 43:183–191

Chan JR, Jolicoeur C, Yamauchi J, Elliott J, Fawcett JP, Ng BK, Cayouette M (2006) The polarity protein Par-3 directly interacts with p75NTR to regulate myelination. Science 314:832–836

Chang IA, Oh MJ, Kim MH, Park SK, Kim BG, Namgung U (2012) Vimentin phosphorylation by Cdc2 in Schwann cell controls axon growth via beta1-integrin activation. FASEB J 26:2401–2413

Chen LM, Bailey D, Fernandez-Valle C (2000) Association of beta 1 integrin with focal adhesion kinase and paxillin in differentiating Schwann cells. J Neurosci 20:3776–3784

Cheng HL, Steinway ML, Russell JW, Feldman EL (2000) GTPases and phosphatidylinositol 3-kinase are critical for insulin-like growth factor-I-mediated Schwann cell motility. J Biol Chem 275:27197–27204

Citri A, Skaria KB, Yarden Y (2003) The deaf and the dumb: the biology of ErbB-2 and ErbB-3. Exp Cell Res 284:54–65

Cosgaya JM, Shooter EM (2001) Binding of nerve growth factor to its p75 receptor in stressed cells induces selective IkappaB-beta degradation and NF-kappaB nuclear translocation. J Neurochem 79:391–399

Cosgaya JM, Chan JR, Shooter EM (2002) The neurotrophin receptor p75NTR as a positive modulator of myelination. Science 298:1245–1248

D'Antonio M, Michalovich D, Paterson M, Droggiti A, Woodhoo A, Mirsky R, Jessen KR (2006) Gene profiling and bioinformatic analysis of Schwann cell embryonic development and myelination. Glia 53:501–515

Delague V, Jacquier A, Hamadouche T, Poitelon Y, Baudot C, Boccaccio I, Chouery E, Chaouch M, Kassouri N, Jabbour R, Grid D, Mégarbané A, Haase G, Lévy N (2007) Mutations in FGD4

encoding the Rho GDP/GTP exchange factor FRABIN cause autosomal recessive Charcot-Marie-Tooth type 4H. Am J Hum Genet 81:1–16
Donaldson JG, Jackson CL (2011) ARF family G proteins and their regulators: roles in membrane transport, development and disease. Nat Rev Mol Cell Biol 12:362–375
D'Souza-Schorey C, Chavrier P (2006) ARF proteins: roles in membrane traffic and beyond. Nat Rev Mol Cell Biol 7:347–358
Feltri ML, Graus Porta D, Previtali SC, Nodari A, Migliavacca B, Cassetti A, Littlewood-Evans A, Reichardt LF, Messing A, Quattrini A, Mueller U, Wrabetz L (2001) Conditional disruption of beta 1 integrin in Schwann cells impedes interactions with axons. J Cell Biol 156:199–209
Fleck D, van Bebber F, Colombo A, Galante C, Schwenk BM, Rabe L, Hampel H, Novak B, Kremmer E, Tahirovic S, Edbauer D, Lichtenthaler SF, Schmid B, Willem M, Haass C (2013) Dual cleavage of neuregulin 1 Type III by BACE1 and ADAM17 liberates its EGF-like domain and allows paracrine signaling. J Neurosci 33:7856–7869
Garratt AN, Britsch S, Birchmeier C (2000) Neuregulin, a factor with many functions in the life of a Schwann cell. Bioessays 22:987–996
Geiger C, Nagel W, Boehm T, van Kooyk Y, Figdor CG, Kremmer E, Hogg N, Zeitlmann L, Dierks H, Weber KS, Kolanus W (2000) Cytohesin-1 regulates beta2 integrin-mediated adhesion through both ARF-GEF function and interaction with LFA-1. EMBO J 19:2525–2536
Gloerich M, Bos JL (2010) Epac: defining a new mechanism for cAMP action. Annu Rev Pharmacol Toxicol 50:355–575
Grossmann KS, Wende H, Paul FE, Cheret C, Garratt AN, Zurborg S, Feinberg K, Besser D, Schulz H, Peles E, Selbach M, Birchmeier W, Birchmeier C (2009) The tyrosine phosphatase Shp2 (PTPN11) directs Neuregulin-1/ErbB signaling throughout Schwann cell development. Proc Natl Acad Sci USA 106:16704–16709
Grove M, Komiyama NH, Nave KA, Grant SG, Sherman DL, Brophy PJ (2007) FAK is required for axonal sorting by Schwann cells. J Cell Biol 176:277–282
Heermann S, Spittau B, Zajzon K, Schwab MH, Krieglstein K (2012) Schwann cells migrate along axons in the absence of GDNF signaling. BMC Neurosci 13:92
Heinen A, Kremer D, Göttle P, Kruse F, Hasse B, Lehmann H, Hartung HP, Küry P (2008) The cyclin-dependent kinase inhibitor p57kip2 is a negative regulator of Schwann cell differentiation and in vitro myelination. Proc Natl Acad Sci USA 105:8748–8753
Hossain S, Fragoso G, Mushynski WE, Almazan G (2010) Regulation of peripheral myelination by Src-like kinases. Exp Neurol 226:47–57
Hu X, He W, Diaconu C, Tang X, Kidd GJ, Macklin WB, Trapp BD, Yan R (2008) Genetic deletion of BACE1 in mice affects remyelination of sciatic nerves. FASEB J 22:2970–2980
Huang C, Jacobson K, Schaller MD (2004) J Cell Sci 117:4619–4628
Huettl RE, Huber AB (2011) Cranial nerve fasciculation and Schwann cell migration are impaired after loss of Npn-1. Dev Biol 359:230–241
Hunter T (2009) Tyrosine phosphorylation: thirty years and counting. Curr Opin Cell Biol 21: 140–146
Inoue G, Gaultier A, Li X, Mantuano E, Richardson G, Talnoue G, Gaultier A, Li X, Mantuano E, Richardson G, Takahashi K, Campana WM, Kahashi K, Campana WM (2010) Erythropoietin promotes Schwann cell migration and assembly of the provisional extracellular matrix by recruiting beta1 integrin to the cell surface. Glia 58:399–409
Ishii A, Furusho M, Bansal R (2013) Sustained activation of ERK1/2 MAPK in oligodendrocytes and Schwann cells enhances myelin growth and stimulates oligodendrocyte progenitor expansion. J Neurosci 33:175–186
Iwase T, Jung CG, Bae H, Zhang M, Soliven B (2005) Glial cell line-derived neurotrophic factor-induced signaling in Schwann cells. J Neurochem 94:1488–1499
Jessen KR, Mirsky R (2005) The origin and development of glial cells in peripheral nerves. Nat Rev Neurosci 6:671–682
Jin F, Dong B, Georgiou J, Jiang Q, Zhang J, Bharioke A, Qiu F, Lommel S, Feltri ML, Wrabetz L, Roder JC, Eyer J, Chen X, Peterson AC, Siminovitch KA (2011) N-WASp is required for

Schwann cell cytoskeletal dynamics, normal myelin gene expression and peripheral nerve myelination. Development (Camb) 138:1329–1337

Kahn RA, Cherfils J, Elias M, Lovering RC, Munro S, Schurmann A (2006) Nomenclature for the human Arf family of GTP-binding proteins: ARF, ARL, and SAR proteins. J Cell Biol 172:645–650

Kolanus W, Nagel W, Schiller B, Zeitlmann L, Godar S, Stockinger H, Seed B (1996) Alpha L beta 2 integrin/LFA-1 binding to ICAM-1 induced by cytohesin-1, a cytoplasmic regulatory molecule. Cell 86:233–242

La Marca R, Cerri F, Horiuchi K, Bachi A, Feltri ML, Wrabetz L, Blobel CP, Quattrini A, Salzer JL, Taveggia C (2011) TACE (ADAM17) inhibits Schwann cell myelination. Nat Neurosci 14:857–865

Lee S, Leach MK, Redmond SA, Chong SY, Mellon SH, Tuck SJ, Feng ZQ, Corey JM, Chan JR (2012) A culture system to study oligodendrocyte myelination processes using engineered nanofibers. Nat Methods 9:917–922

Lee S, Chong SY, Tuck SJ, Corey JM, Chan JR (2013) A rapid and reproducible assay for modeling myelination by oligodendrocytes using engineered nanofibers. Nat Protoc 8:771–782

Lemke G (2006) Neuregulin-1 and myelination. Sci STKE 2006:pe11

Li R, Chen J, Hammonds G, Phillips H, Armanini M, Wood P, Bunge R, Godowski PJ, Sliwkowski MX, Mather JP (1996) Identification of Gas6 as a growth factor for human Schwann cells. J Neurosci 16:2012–2019

Liu L, Pohajdak B (1992) Cloning and sequencing of a human cDNA from cytolytic NK/T cells with homology to yeast SEC7. Biochim Biophys Acta 1132:75–78

Lyons DA, Pogoda HM, Voas MG, Woods IG, Diamond B, Nix R, Arana N, Jacobs J, Talbot WS (2005) ErbB3 and ErbB2 are essential for Schwann cell migration and myelination in zebrafish. Curr Biol 15:513–524

Macklin WB (2010) The myelin brake: when enough is enough. Sci Signal 3:pe32

Mahanthappa NK, Anton ES, Matthew WD (1996) Glial growth factor 2, a soluble neuregulin, directly increases Schwann cell motility and indirectly promotes neurite outgrowth. J Neurosci 16:4673–4683

Mantuano E, Jo M, Gonias SL, Campana WM (2010) Low density lipoprotein receptor-related protein (LRP1) regulates Rac1 and RhoA reciprocally to control Schwann cell adhesion and migration. J Biol Chem 285:14259–14266

Mantuano E, Henry K, Yamauchi T, Hiramatsu N, Yamauchi K, Orita S, Takahashi K, Lin JH, Gonias SL, Campana WM (2011) The unfolded protein response is a major mechanism by which LRP1 regulates Schwann cell survival after injury. J Neurosci 31:13376–13385

Masaki T, Qu J, Cholewa-Waclaw J, Burr K, Raaum R, Rambukkana A (2013) Reprogramming adult Schwann cells to stem cell-like cells by leprosy bacilli promotes dissemination of infection. Cell 152:51–67

Matsuda M, Kurata T (1996) Emerging components of the Crk oncogene product: the first identified adaptor protein. Cell Signal 8:335–440

Meacci E, Tsai SC, Adamik R, Moss J, Vaughan M (1997) Cytohesin-1, a cytosolic guanine nucleotide-exchange protein for ADP-ribosylation factor. Proc Natl Acad Sci USA 94:1745–1748

Melendez-Vasquez CV, Einheber S, Salzer JL (2004) Rho kinase regulates Schwann cell myelination and formation of associated axonal domains. J Neurosci 24:3953–3963

Miyamoto Y, Yamauchi J (2010) Cellular signaling of Dock family proteins in neural signaling. Cell Signal 22:175–182

Miyamoto Y, Yamauchi J, Tanoue A, Wu C, Mobley WC (2006) TrkB binds and tyrosine-phosphorylates Tiam1, leading to activation of Rac1 and induction of changes in cellular morphology. Proc Natl Acad Sci USA 103:10444–10449

Miyamoto Y, Torii T, Yamamori N, Eguchi T, Nagao M, Nakamura K, Tanoue A, Yamauchi J (2012) Paxillin is the target of c-Jun N-terminal kinase in Schwann cells and regulates migration. Cell Signal 24:2061–2069

Miyamoto Y, Torii T, Nakamura K, Takashima S, Sanbe A, Tanoue A, Yamauchi J (2013) Signaling through Arf6 guanine-nucleotide exchange factor cytohesin-1 regulates migration in Schwann cells. Cell Signal 25:1379–1387

Monje PV, Rendon S, Athauda G, Bates M, Wood PM, Bunge MB (2009) Non-antagonistic relationship between mitogenic factors and cAMP in adult Schwann cell re-differentiation. Glia 57:947–961

Monk KR, Naylor SG, Glenn TD, Mercurio S, Perlin JR, Dominguez C, Moens CB, Talbot WS (2009) A G protein-coupled receptor is essential for Schwann cells to initiate myelination. Science 325:1402–1405

Monk KR, Oshima K, Jörs S, Heller S, Talbot WS (2011) Gpr126 is essential for peripheral nerve development and myelination in mammals. Development (Camb) 138:2673–2680

Morris JK, Lin W, Hauser C, Marchuk Y, Getman D, Lee KF (1999) Rescue of the cardiac defect in ErbB2 mutant mice reveals essential roles of ErbB2 in peripheral nervous system development. Neuron 23:273–283

Najm FJ, Lager AM, Zaremba A, Wyatt K, Caprariello AV, Factor DC, Karl RT, Maeda T, Miller RH, Tesar PJ (2013) Transcription factor–mediated reprogramming of fibroblasts to expandable, myelinogenic oligodendrocyte progenitor cells. Nat Biotechnol 31(5):426–433. doi:10.1038/nbt.2561

Napoli I, Noon LA, Ribeiro S, Kerai AP, Parrinello S, Rosenberg LH, Collins MJ, Harrisingh MC, White IJ, Woodhoo A, Lloyd AC (2012) A central role for the ERK-signaling pathway in controlling Schwann cell plasticity and peripheral nerve regeneration in vivo. Neuron 73: 729–742

Nave KA, Salzer JL (2006) Axonal regulation of myelination by neuregulin 1. Curr Opin Neurobiol 16:492–500

Newbern JM, Li X, Shoemaker SE, Zhou J, Zhong J, Wu Y, Bonder D, Hollenback S, Coppola G, Geschwind DH, Landreth GE, Snider WD (2011) Specific functions for ERK/MAPK signaling during PNS development. Neuron 69:91–105

Ng AA, Logan AM, Schmidt EJ, Robinson FL (2013) The CMT4B disease-causing phosphatases Mtmr2 and Mtmr13 localize to the Schwann cell cytoplasm and endomembrane compartments, where they depend upon each other to achieve wild-type levels of protein expression. Hum Mol Genet 22:1493–1506

Nguyen KQ, Tsou WI, Kotenko S, Birge RB (2013) TAM receptors in apoptotic cell clearance, autoimmunity, and cancer. Autoimmunity 46(5):294–297. doi:10.3109/08916934.2013.794515

Novak N, Bar V, Sabanay H, Frechter S, Jaegle M, Snapper SB, Meijer D, Peles E (2011) N-WASP is required for membrane wrapping and myelination by Schwann cells. J Cell Biol 192: 243–250

Nodari A, Zambroni D, Quattrini A, Court FA, D'Urso A, Recchia A, Tybulewicz VL, Wrabetz L, Feltri ML (2007) Beta1 integrin activates Rac1 in Schwann cells to generate radial lamellae during axonal sorting and myelination. J Cell Biol 177:1063–1075

Ogata T, Iijima S, Hoshikawa S, Miura T, Yamamoto S, Oda H, Nakamura K, Tanaka S (2004) Opposing extracellular signal-regulated kinase and Akt pathways control Schwann cell myelination. J Neurosci 24:6724–6732

Ogata T, Yamamoto S, Nakamura K, Tanaka S (2006) Signaling axis in Schwann cell proliferation and differentiation. Mol Neurobiol 33:51–62

Orita S, Henry K, Mantuano E, Yamauchi K, De Corato A, Ishikawa T, Feltri ML, Wrabetz L, Gaultier A, Pollack M, Ellisman M, Takahashi K, Gonias SL, Campana WM (2013) Schwann cell LRP1 regulates remak bundle ultrastructure and axonal interactions to prevent neuropathic pain. J Neurosci 33:5590–5602

Paratcha G, Ledda F, Ibáñez CF (2003) The neural cell adhesion molecule NCAM is an alternative signaling receptor for GDNF family ligands. Cell 113:867–879

Parkinson DB, Bhaskaran A, Droggiti A, Dickinson S, D'Antonio M, Mirsky R, Jessen KR (2004) Krox-20 inhibits Jun-NH2-terminal kinase/c-Jun to control Schwann cell proliferation and death. J Cell Biol 164:385–394

Pereira JA, Lebrun-Julien F, Suter U (2012) Molecular mechanisms regulating myelination in the peripheral nervous system. Trends Neurosci 35:123–134

Perlin JR, Lush ME, Stephens WZ, Piotrowski T, Talbot WS (2011) Neuronal neuregulin 1 type III directs Schwann cell migration. Development (Camb) 138:4639–4648

Pogoda HM, Sternheim N, Lyons DA, Diamond B, Hawkins TA, Woods IG, Bhatt DH, Franzini-Armstrong C, Dominguez C, Arana N, Jacobs J, Nix R, Fetcho JR, Talbot WS (2006) A genetic screen identifies genes essential for development of myelinated axons in zebrafish. Dev Biol 298:118–131

Raphael AR, Talbot WS (2011) New insights into signaling during myelination in zebrafish. Curr Top Dev Biol 97:1–19

Rangaraju S, Verrier JD, Madorsky I, Nicks J, Dunn WA Jr, Notterpek L (2010) Rapamycin activates autophagy and improves myelination in explant cultures from neuropathic mice. J Neurosci 30:11388–11397

Reddien PW, Horvitz HR (2004) The engulfment process of programmed cell death in *Caenorhabditis elegans*. Annu Rev Cell Dev Biol 20:193–221

Riethmacher D, Sonnenberg-Riethmacher E, Brinkmann V, Yamaai T, Lewin GR, Birchmeier C (1997) Severe neuropathies in mice with targeted mutations in the ErbB3 receptor. Nature (Lond) 389:725–730

Rossman KL, Der CJ, Sondek J (2005) GEF means go: turning on RHO GTPases with guanine nucleotide-exchange factors. Nat Rev Mol Cell Biol 6:167–180

Scherer SS (2001) Myelination: some receptors required. J Cell Biol 156:13–15

Schmidt A, Hall A (2002) Guanine nucleotide exchange factors for Rho GTPases: turning on the switch. Genes Dev 16:1587–1609

Shi F, Telesco SE, Liu Y, Radhakrishnan R, Lemmon MA (2010) ErbB3/HER3 intracellular doErbB3/HER3 intracellular domain is competent to bind ATP and catalyze autophosphorylation main is competent to bind ATP. Proc Natl Acad Sci USA 107:7692–7697

Sparrow N, Manetti ME, Bott M, Fabianac T, Petrilli A, Bates ML, Bunge MB, Lambert S, Fernandez-Valle C (2012) The actin-severing protein cofilin is downstream of neuregulin signaling and is essential for Schwann cell myelination. J Neurosci 32:5284–5297

Stendel C, Roos A, Deconinck T, Pereira J, Castagner F, Niemann A, Kirschner J, Korinthenberg R, Ketelsen UP, Battaloglu E, Parman Y, Nicholson G, Ouvrier R, Seeger J, De Jonghe P, Weis J, Krüttgen A, Rudnik-Schöneborn S, Bergmann C, Suter U, Zerres K, Timmerman V, Relvas JB, Senderek J (2007) Peripheral nerve demyelination caused by a mutant Rho GTPase guanine nucleotide exchange factor, frabin/FGD4. Am J Hum Genet 81:158–164

Syed N, Reddy K, Yang DP, Taveggia C, Salzer JL, Maurel P, Kim HA (2010) Soluble neuregulin-1 has bifunctional, concentration-dependent effects on Schwann cell myelination. J Neurosci 30:6122–6131

Taveggia C, Feltri ML, Wrabetz L (2010) Signals to promote myelin formation and repair. Nat Rev Neurol 6:276–287

Tep C, Kim ML, Opincariu LI, Limpert AS, Chan JR, Appel B, Carter BD, Yoon SO (2012) Brain-derived neurotrophic factor (BDNF) induces polarized signaling of small GTPase (Rac1) protein at the onset of Schwann cell myelination through partitioning-defective 3 (Par3) protein. J Biol Chem 287:1600–1608

Triolo D, Dina G, Taveggia C, Vaccari I, Porrello E, Rivellini C, Domi T, La Marca R, Cerri F, Bolino A, Quattrini A, Previtali SC (2012) Vimentin regulates peripheral nerve myelination. Development (Camb) 39:1359–1367

Turner CE (2000) Paxillin interactions. J Cell Sci 113:4139–4140

Verhoeven K, De Jonghe P, Van de Putte T, Nelis E, Zwijsen A, Verpoorten N, De Vriendt E, Jacobs A, Van Gerwen V, Francis A, Ceuterick C, Huylebroeck D, Timmerman V (2003) Slowed conduction and thin myelination of peripheral nerves associated with mutant rho guanine-nucleotide exchange factor 10. Am J Hum Genet 73:926–932

Willem M, Garratt AN, Novak B, Citron M, Kaufmann S, Rittger A, DeStrooper B, Saftig P, Birchmeier C, Haass C (2006) Control of peripheral nerve myelination by the beta-secretase BACE1. Science 314:664–666

Woldeyesus MT, Britsch S, Riethmacher D, Xu L, Sonnenberg-Riethmacher E, Abou-Rebyeh F, Harvey R, Caroni P, Birchmeier C (1999) Peripheral nervous system defects in ErbB2 mutants following genetic rescue of heart development. Genes Dev 13:2538–2548

Yamauchi J, Chan JR, Shooter EM (2003) Neurotrophin 3 activation of TrkC induces Schwann cell migration through the c-Jun N-terminal kinase pathway. Proc Natl Acad Sci USA 100:14421–14426

Yamauchi J, Chan JR, Shooter EM (2004) Neurotrophins regulate Schwann cell migration by activating divergent signaling pathways dependent on Rho GTPases. Proc Natl Acad Sci USA 101:8774–8779

Yamauchi J, Chan JR, Miyamoto Y, Tsujimoto G, Shooter EM (2005a) The neurotrophin-3 receptor TrkC directly phosphorylates and activates the nucleotide exchange factor Dbs to enhance Schwann cell migration. Proc Natl Acad Sci USA 102:5198–5203

Yamauchi J, Miyamoto Y, Tanoue A, Shooter EM, Chan JR (2005b) Ras activation of a Rac1 exchange factor, Tiam1, mediates neurotrophin-3-induced Schwann cell migration. Proc Natl Acad Sci USA 102:14889–14894

Yamauchi J, Miyamoto Y, Sanbe A, Tanoue A (2006) JNK phosphorylation of paxillin, acting through the Rac1 and Cdc42 signaling cascade, mediates neurite extension in N1E-115 cells. Exp Cell Res 312:2954–2961

Yamauchi J, Miyamoto Y, Murabe M, Fujiwara Y, Sanbe A, Fujita Y, Murase S, Tanoue A (2007) Gadd45a, the gene induced by the mood stabilizer valproic acid, regulates neurite outgrowth through JNK and the substrate paxillin in N1E-115 neuroblastoma cells. Exp Cell Res 313:1886–1896

Yamauchi J, Miyamoto Y, Kusakawa S, Torii T, Mizutani R, Sanbe A, Nakajima H, Kiyokawa N, Tanoue A (2008a) Neurofibromatosis 2 tumor suppressor, the gene induced by valproic acid, mediates neurite outgrowth through interaction with paxillin. Exp Cell Res 314:2279–2288

Yamauchi J, Miyamoto Y, Chan JR, Tanoue A (2008b) ErbB2 directly activates the exchange factor Dock7 to promote Schwann cell migration. J Cell Biol 181:351–365

Yamauchi J, Miyamoto Y, Hamasaki H, Sanbe A, Kusakawa S, Nakamura A, Tsumura H, Maeda M, Nemoto N, Kawahara K, Torii T, Tanoue A (2011) The atypical guanine-nucleotide exchange factor, Dock7, negatively regulates Schwann cell differentiation and myelination. J Neurosci 31:12579–12592

Yamauchi J, Miyamoto Y, Torii T, Takashima S, Kondo K, Kawahara K, Nemoto N, Chan JR, Tsujimoto G, Tanoue A (2012) Phosphorylation of cytohesin-1 by Fyn is required for initiation of myelination and the extent of myelination during development. Sci Signal 5:ra69

Yang DP, Kim J, Syed N, Tung YJ, Bhaskaran A, Mindos T, Mirsky R, Jessen KR, Maurel P, Parkinson DB, Kim HA (2012) p38 MAPK activation promotes denervated Schwann cell phenotype and functions as a negative regulator of Schwann cell differentiation and myelination. J Neurosci 32:7158–7168

Yang N, Zuchero JB, Ahlenius H, Marro S, Ng YH, Vierbuchen T, Hawkins JS, Geissler R, Barres BA, Wernig M (2013) Generation of oligodendroglial cells by direct lineage conversion. Nat Biotechnol 31(5):434–439. doi:10.1038/nbt.2564

Zhou FQ, Zhong J, Snider WD (2003) Extracellular crosstalk: when GDNF meets N-CAM. Cell 113:814–815

Chapter 3
Membrane Skeleton in Schmidt–Lanterman Incisure in Schwann Cells of the Peripheral Nervous System

Nobuo Terada, Yurika Saitoh, Nobuhiko Ohno, and Shinichi Ohno

Abstract Schmidt–Lanterman incisure (SLI), a truncated cone–shape in a myelin internode, is a specific feature of myelinated nerve fibers in the peripheral nervous system (PNS). In this review, we focus on the membrane skeleton in SLI. First, we describe a membrane skeletal protein, 4.1G, and its relationship to membrane palmitoylated protein 6 (MPP6) and cell adhesion molecule 4 (CADM4), which is analogous to a molecular complex in the erythrocyte membrane skeleton, 4.1R–MPP1–glycophorin C. In 4.1G-deficient nerve fibers, the height of the SLI-truncated cones was reduced compared to that in the wild type. 4.1G was essential for molecular targeting of MPP6 and CADM4 in SLI. Second, we discuss a signal transduction protein, Src, in the SLIs of mouse sciatic nerves, and its phosphorylation states under normal conditions or deletion of 4.1G. Normally, Src is phosphorylated in Y527, but not in Y418. Developmentally, the phosphorylation in Y418 appeared in SLIs of early postnatal mouse sciatic nerves. An MPP6–Src interaction was found, and the phosphorylation of Y418 appeared in 4.1G-deficient nerve fibers. The functional meaning of the Src localization in SLI is discussed. Here, we demonstrate a novel Src–MPP6–4.1G–CADM4 membrane skeletal molecular complex in SLIs, with potential roles in regulation of adhesion and signal transduction in Schwann cells.

Keywords Cell adhesion molecule • Membrane skeleton • Peripheral nervous system • Protein 4.1 • Schmidt–Lanterman incisure • Schwann cell • Src

N. Terada (✉)
Department of Occupational Therapy, School of Health Sciences, Shinshu University School of Medicine, 3-1-1 Asahi, Matsumoto, Nagano 390-8621, Japan

Department of Anatomy and Molecular Histology, Interdisciplinary Graduate School of Medicine and Engineering, University of Yamanashi, Chuo City, Yamanashi, Japan
e-mail: nobuot@shinshu-u.ac.jp

Y. Saitoh • N. Ohno • S. Ohno
Department of Anatomy and Molecular Histology, Interdisciplinary Graduate School of Medicine and Engineering, University of Yamanashi, Chuo City, Yamanashi, Japan

K. Sango and J. Yamauchi (eds.), *Schwann Cell Development and Pathology*,
DOI 10.1007/978-4-431-54764-8_3, © Springer Japan 2014

3.1 Membrane Skeleton and Protein 4.1 Families

Membrane skeleton is a term for a fine meshwork structure under the cell membranes that was originally identified in erythrocytes (Da Costa et al. 2013). It was observed three dimensionally by transmission electron microscopy with a quick-freezing and deep-etching method after erythrocyte splitting as demonstrated in Fig. 3.1a–c (Ohno et al. 1994; Terada et al. 1997). The proteins forming this membrane skeletal structure in erythrocytes have been well documented (Da Costa et al. 2013). Spectrin–actin networks are connected with ankyrin and protein 4.1R (4.1R), which connect to intramembranous proteins, band 3 (anion exchanger), and glycophorin C (Machnicka et al. 2013) (Fig. 3.1). Gradually, it was proved that the

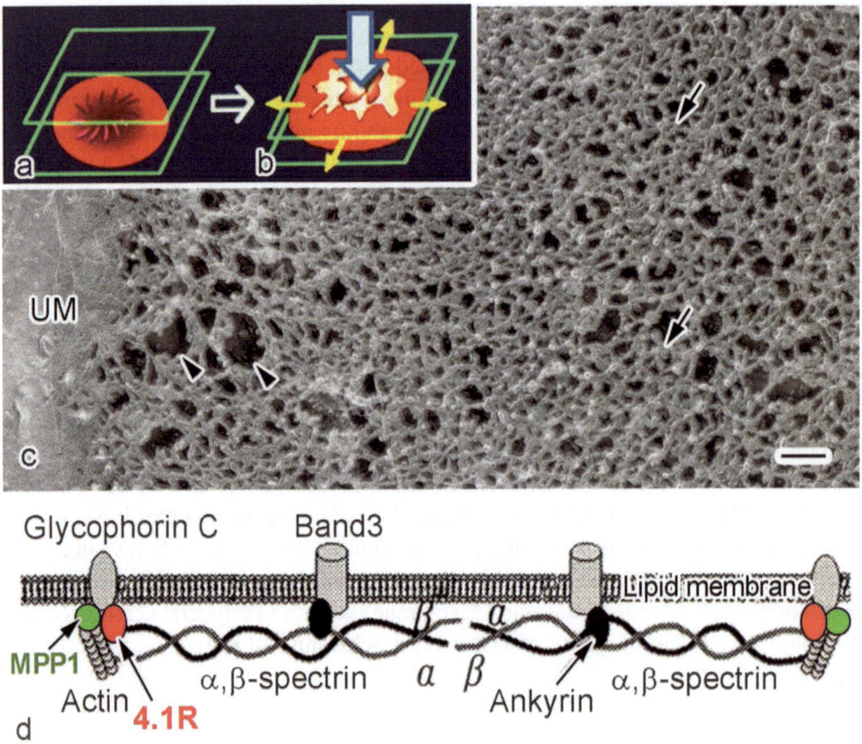

Fig. 3.1 Procedures for erythrocyte splitting (**a, b**) and a transmission electron micrograph of membrane skeletons in human erythrocytes using a quick-freezing and deep-etching method (**c**). Briefly, erythrocytes were sandwiched between silane-glutaraldehyde-coated coverslips (**a**) and mildly pressed to become attached to them (**b**). Some parts of the erythrocyte lipid membranes are peeled off, and fine meshwork structures are observed (*arrows* in **c**) under the upper lipid membrane (*UM* in **c**). Some parts of the meshwork are teased to make large artificial holes (*arrowheads* in **c**), probably because of strong mechanical forces. A schematic image (**d**) depicts molecular components of membrane skeletons in erythrocytes. Note the 4.1R–MPP1–glycophorin C molecular complex at one side of the spectrin network, which interacts with actin filaments. *Bar* 200 nm

membrane skeletal components are localized not only in erythrocytes but also in many tissues and organs for specific functions such as adhesion (Baines 2010).

There are four 4.1-family members (Peters et al. 1998). The 4.1 families have three domains; an ezrin-radixin-moesin (ERM) domain, a spectrin–actin-binding domain, and a C-terminal domain. The ERM domain has a role in interaction with several intramembranous proteins (Baines 2010). 4.1G is a family of 4.1R (Parra et al. 1998; Walensky et al. 1998), which links the spectrin–actin network (Discher et al. 1995). In this review, we mainly focus on the 4.1G-including molecular complex in the peripheral nervous system (PNS), based on our finding that 4.1G is present in rodent Schwann cells (Chen et al. 2011; Ohno et al. 2006).

3.2 Structure of Schmidt–Lanterman Incisure (SLI)

SLI is a characteristic structure of Schwann cell membranes in the internodes of myelinated nerve fibers in the PNS. It is a funnel-shaped interruption within a myelin sheath of nerve fibers, and its dynamic structure is observed with an "in vivo cryotechnique" (IVCT) (Fig. 3.2a, b). For electron microscopy, IVCT was performed by a combination of cutting sciatic nerves with a fine needle and simultaneous pouring of isopentane-propane cryogen (−193 °C) precooled in liquid nitrogen (Ohno et al. 1996). At the light microscopic level, pouring the isopentane-propane cryogen alone is sufficient because of lower resolution compared to the electron microscopic images (Ohno et al. 2010; Saitoh et al. 2012; Terada et al. 2012b) (Fig. 3.3). The SLI has three specific structural features, as illustrated in Fig. 3.2c: (1) a coiled spring shape from their cytoplasmic continuity, (2) adhesion of membranes with adjacent membranes, and (3) swelling of cytoplasm that differs from the tight interaction of adjacent membranes in myelin parts.

Elasticity and stability of the adhesion system and membrane skeleton in SLI are probably important for maintenance of the PNS internodes. Concerning molecules inside SLIs, high concentrations of actin and spectrin, which are the main components of the membrane skeleton, have been reported (Susuki et al. 2011; Trapp et al. 1989). Concerning adhesion structures in SLI, various tight and adherens junctional molecules have also been identified (Poliak et al. 2002), which include claudin (Poliak et al. 2002), occludin (Alanne et al. 2009), E-cadherin (Tricaud et al. 2005; Young et al. 2002), and junctional adhesion molecule (JAM)-C (Scheiermann et al. 2007). Additionally, we have already reported membrane skeletal proteins and adhesion molecules as components of SLI, such as 4.1G (Ohno et al. 2006) (Fig. 3.3), a membrane-associated guanylate kinase (MAGUK) protein, membrane protein palmytoylated 6 (MPP6) (Terada et al. 2012a), and an intramembranous adhesion molecule, cell adhesion molecule 4 (CADM4) (Ivanovic et al. 2012).

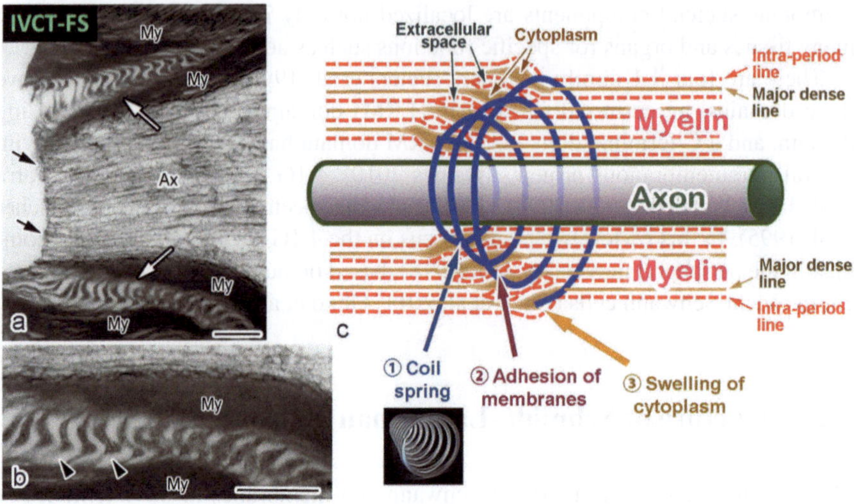

Fig. 3.2 Transmission electron micrographs of a Schmidt–Lanterman incisure (SLI) in a mouse sciatic nerve fiber with an in vivo cryotechnique (*IVCT*) followed by freeze-substitution fixation (FS) (**a**, **b**) and a schematic model structure of SLI (**c**). The living mouse sciatic nerve fiber was cryocut with a knife precooled with liquid nitrogen (−196 °C), and isopentane-propane cryogen (−193 °C) was poured on it, followed by cooling with liquid nitrogen (**a**, **b**). *Black arrows* in **a** indicate the cryocut tissue surface; *white arrows* also indicate SLI. *Ax* axon, *My* myelin. **b** Highly magnified view of a part of SLI shown in **a**. Within the cytoplasm, some extracellular spaces outside cell membranes (*arrowheads* in **b**) are observed. **c** The three arrows in the schematic illustration of an SLI indicate there are three structural features of SLI: ① a coil spring (an artificial iron-coil spring is shown), ② adhesion of membranes, and ③ swelling of cytoplasm. *Bars* 500 nm

3.3 Protein 4.1G and MAGUK Proteins for Cell–Cell Adhesion

MAGUK family proteins contain PDZ [PSD (postsynaptic density)-95/Dlg (Drosophila disks large)/ZO (zonula occludens)-1], GUK (guanylate kinase) and SH3 (src-homology-3) domains, which localize at specific domains of cell membranes (de Mendoza et al. 2010; Funke et al. 2005). For example, some MAGUKs, such as Dlg and ZO-1, are required for the formation of adherens and tight junctions in epithelial cells. Their function is thought to be as scaffold membranes (Lozovatsky et al. 2009; Nix et al. 2000). In addition, several MAGUKs were also found to control intracellular protein transport through their ability to bind motor proteins (Verhey and Rapoport 2001; Yamada et al. 2007). Some MAGUKs contain specific domains that interact with 4.1 proteins (Lue et al. 1994, 1996).

From these MAGUK families, we found that MPP6 (also designated Pals2, VAM1, and p55T; http://www.genenames.org/) was a novel MAGUK molecule that interacts with 4.1G in peripheral nerve fibers, especially in SLIs (Terada et al. 2012a). To examine the functional relationship of 4.1G and MPP6, we used

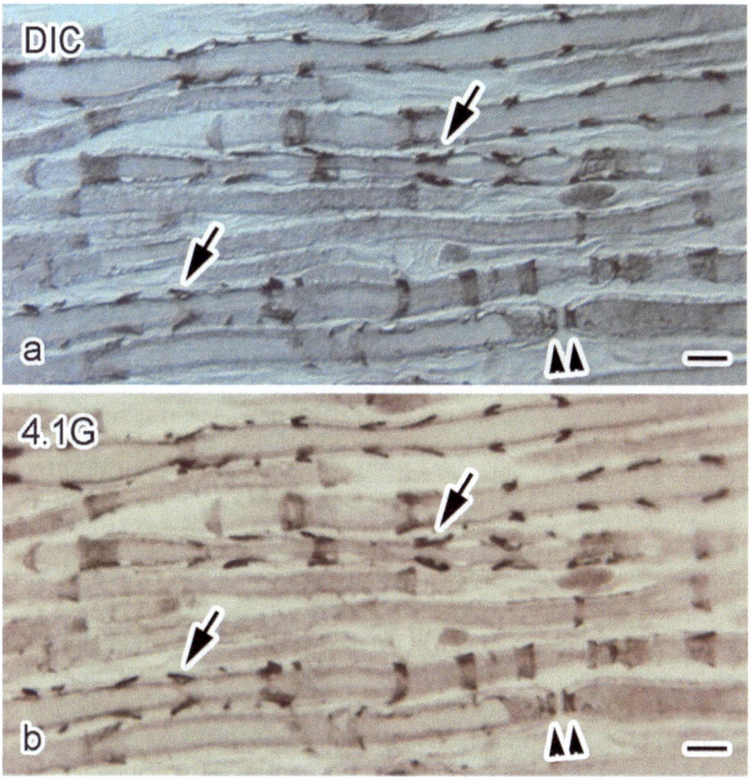

Fig. 3.3 Diffraction interference contrast (*DIC*) image (**a**) and a *4.1G*-immunostaining image (**b**) of the same mouse sciatic nerve fiber using an "in vivo cryotechnique" followed by freeze-substitution fixation. In this case, liquid isopentane-propane cryogen (−193 °C) cooled with liquid nitrogen was directly poured over the specimens. The nerve fiber structures reflect their living states using this cryotechnique. By 4.1G immunostaining (**b**), funnel-shaped structures of SLIs are clearly detected in the internodes (*arrows*). Various heights are seen along myelin sheaths. *Double arrowheads* indicate paranodes beside the node of Ranvier. *Bars* 10 μm

4.1G-deficient mouse (–/–) peripheral nerve fibers (Fig. 3.4a, b). In the 4.1G–/– mouse sciatic nerve fibers, MPP6 was rarely localized in SLI and abnormally detected in the cytoplasm near the Schwann cell nuclei (Terada et al. 2012a). In addition to the disruption of MPP6 targeting, the total amount of MPP6 was also significantly reduced. However, the molecular mechanism for the reduction has not been clearly identified. Nevertheless, MPP6 was still present at the paranodal loops even in the 4.1G–/– sciatic nerve fibers, indicating the independency of 4.1G in the MPP6 targeting to paranodes.

Originally, MPP6 was identified in epithelial cells as a mammalian homologue of Lin-7 (mLin-7)-binding protein (Kamberov et al. 2000; Tseng et al. 2001), and is thought to play a role in targeting of proteins to basolateral membrane surfaces. The existence of a MPP6–4.1G complex in mouse Schwann cells is analogous to the

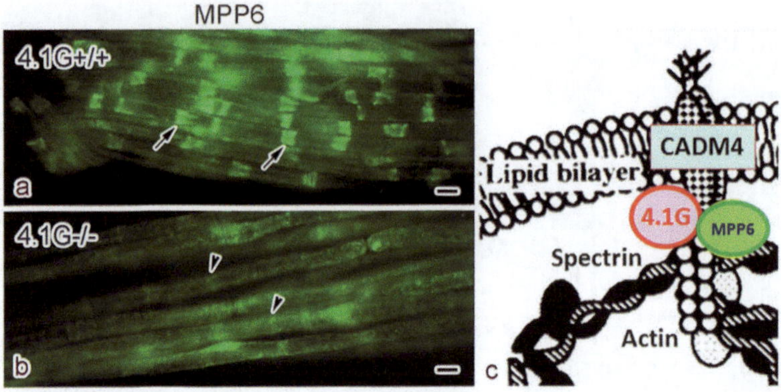

Fig. 3.4 Two MPP6 immunostaining images in 4.1G+/+ (**a**, wild-type) and 4.1G–/– (**b**) mouse sciatic nerve fibers, and a schematic 4.1G–MPP6–CADM4 molecular complex in SLI (**c**). MPP6 localization is clearly shown in SLI of the 4.1G+/+ nerve fibers (*arrows* in **a**), but 4.1G immunostaining is dramatically reduced in 4.1G–/– nerve fibers (*arrowheads* in **b**). Such a reduction was also detected for CADM4 immunostaining (Ivanovic et al. 2012). Although a schematic model of the 4.1G–MPP6–CADM4 molecular complex is shown (**c**), the actual structural interactions between spectrin and actin, and also their visualization, have not been demonstrated. *Bars* 10 μm

three following molecular complexes: (1) the MPP1 (p55)–4.1R complex found in erythrocytes as demonstrated in Fig. 3.1d (Nunomura et al. 2000; Quinn et al. 2009; Seo et al. 2009), (2) CASK–4.1N interaction in neurons (Biederer and Sudhof 2001), and (3) MPP6–4.1B interaction in epithelial cells (Shingai et al. 2003).

Some MAGUK proteins are thought to function as transportation for cargos and interact with motor molecules on microtubules (Verhey and Rapoport 2001). However, MPP6 did not show a transport function for 4.1G. On the other hand, one of the ERM superfamily proteins, merlin, was reported to directly interact with both microtubules (Xu and Gutmann 1998) and kinesin motor proteins (Bensenor et al. 2010). Although 4.1G is a superfamily of the ERM-containing protein, its interaction with microtubules is obscure. In addition, because merlin was also reported to be expressed in Schwann cells (Scherer and Gutmann 1996), a temporal and spatial study for such an ERM protein, merlin, and 4.1G–MPP6 on microtubules will be a further interesting study.

3.4 Protein 4.1G and Cell Adhesion Molecule (CADM) Families

Cell adhesion molecule (CADM) families are Ca^{2+}-independent immunoglobulin-like cell adhesion molecules, which homophilically and heterophilically interact with themselves or with other protein families (Ogita and Takai 2006). For example, CADM1 (also designated SynCAM/TSLC1/SgIgSF/RA175/Necl-2) was reported to have a role in adhesion in germ cells of the seminiferous tubules

(Wakayama et al. 2001). In the testes of CADM1-deficient mice, dysfunction of germ cells has been reported, indicating its indispensable role in spermatogenesis (Fujita et al. 2006; Surace et al. 2006; van der Weyden et al. 2006; Yamada et al. 2006). CADM1 has a molecular domain to bind to the 4.1-family proteins, and the 4.1G interaction to CADM1 was proved in mouse testes (Terada et al. 2010). The CADM1–4.1B membrane complex was also identified in endocrine cells in adrenal medulla (Ohno et al. 2009).

For Schwann cell–axon interaction, some CADM proteins, such as CADM3 (Necl1), CADM1 (Necl2), and CADM4 (Necl4), were demonstrated (Maurel et al. 2007; Spiegel et al. 2007). Axons highly express CADM3 and CADM1, whereas Schwann cells express CADM4 and lower amounts of CADM1 (Maurel et al. 2007). It was proposed that CADM4 is located along the internodes in direct apposition to CADM3 (Spiegel et al. 2007). Additionally, we have demonstrated disruption of the CADM4 localization in Schwann cells, including SLIs in 4.1G−/− peripheral nerve fibers (Ivanovic et al. 2012). Thus, the 4.1G–MPP6–CADM4 molecular complex is involved in SLI (Fig. 3.4c).

3.5 Size Determination of SLI by Protein 4.1G

In old 4.1G−/− mice, the size of SLIs was reduced compared to that in wild-type mice, as revealed by both E-cadherin and actin staining (Terada et al. 2012a) (Fig. 3.5b). One of the MAGUKs, MPP5 (Pals1), was reported to regulate E-cadherin trafficking in mammalian epithelial cells (Wang et al. 2007). On the other hand, another 4.1 family protein, 4.1R, was shown to link to E-cadherin/β-catenin in mouse stomach epithelial cells (Yang et al. 2009). In Schwann cells, the E-cadherin localization was retained even in 4.1G−/− mice. We assume that the reduction in the SLI size in 4.1G−/− nerve fibers may result from the reduced capability of lipid membrane collection in relationship to the 4.1G–MPP6–CADM4 membrane skeletal molecular complex (Fig. 3.5c). The balance between the CADM4-including adhesion molecular complex as mentioned in Sect. 3.6 and E-cadherin-including adhesion molecules is an interesting point, and we also mention a possible related signal molecule, Src, in Sect. 3.8.

In 4.1R-deficient erythrocytes, it is well known that an elliptocytic shape change from the normal biconcave disc occurs, depending on age, which causes gradual instability of functional membrane skeletons against mechanical strength under circulation (Mohandas and Gallagher 2008). Similar to the erythrocyte shape change (Terada et al. 1998a, b), we now believe that the 4.1G–MPP6–CADM4 molecular complex affects maintenance of the SLI structure during aging. Concerning mechanical strength, as the lengths of nerve fibers in the peripheral nervous system (PNS) easily change with mechanical stretching in animal bodies during exercise, it is possible that the SLI present along myelin internodes in the PNS may have a role in protecting peripheral nerve fibers while under mechanical external forces, as demonstrated in Sect. 3.2.

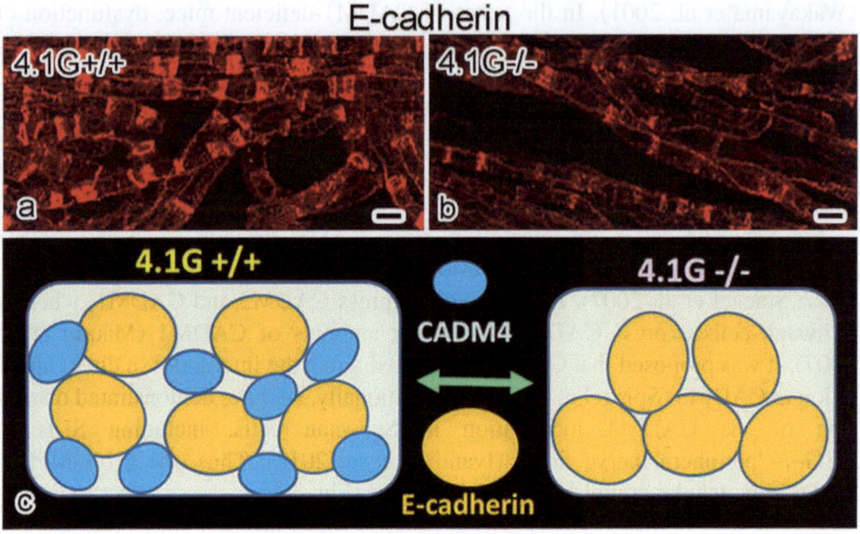

Fig. 3.5 Two E-cadherin immunostaining images in 4.1G+/+ (**a**, wild-type) and 4.1G–/– (**b**) mouse sciatic nerve fibers, and schematic models regarding the relationship between CADM4 and E-cadherin in cell membranes (**c**). Note that E-cadherin immunostaining is detected at SLI even in the 4.1G–/– nerve fibers (**b**), although the heights of SLIs are reduced compared to those in the 4.1G+/+ fibers (**a**). **c** In some areas of the SLI, cell membrane areas may be reduced, accompanied by the loss of CADM4 with the underlying 4.1G–MPP6 membrane skeletal molecular complex. *Bars* 10 μm

3.6 Other MAGUK Proteins in Schwann Cells

Other MAGUKs, such as Discs large 1 (Dlg1) (Bolino et al. 2004; Bolis et al. 2009; Cotter et al. 2010; Goebbels et al. 2010) and MPP5 (Ozcelik et al. 2010), were reported to affect myelination, which is caused by disruption of the phosphatidylinositol metabolism performed by myotubularin-related (MTMR)-2 and phosphatase and tensin homologue (PTEN). On the other hand, it has been demonstrated that MAGUK families interact with each other, such as Dlg1 and MPP7 in epithelial cells (Bohl et al. 2007). Therefore, examination of their appearance and the relationships between MPP6 and MPP5 in Schwann cells is needed.

3.7 Src Kinase Family Protein in SLI

Although the 4.1G–MPP6–CADM4 complex indicates adhesion membrane–skeleton molecular interactions, its relationship to signal transduction inside Schwann cells is unclear. Recently, we have focused on an oncogenic kinase, Src (Terada et al. 2013), because both positive and negative signal roles for cell–cell adhesion were reported in epithelial cells (Shindo et al. 2008). In adult mouse SLIs, we

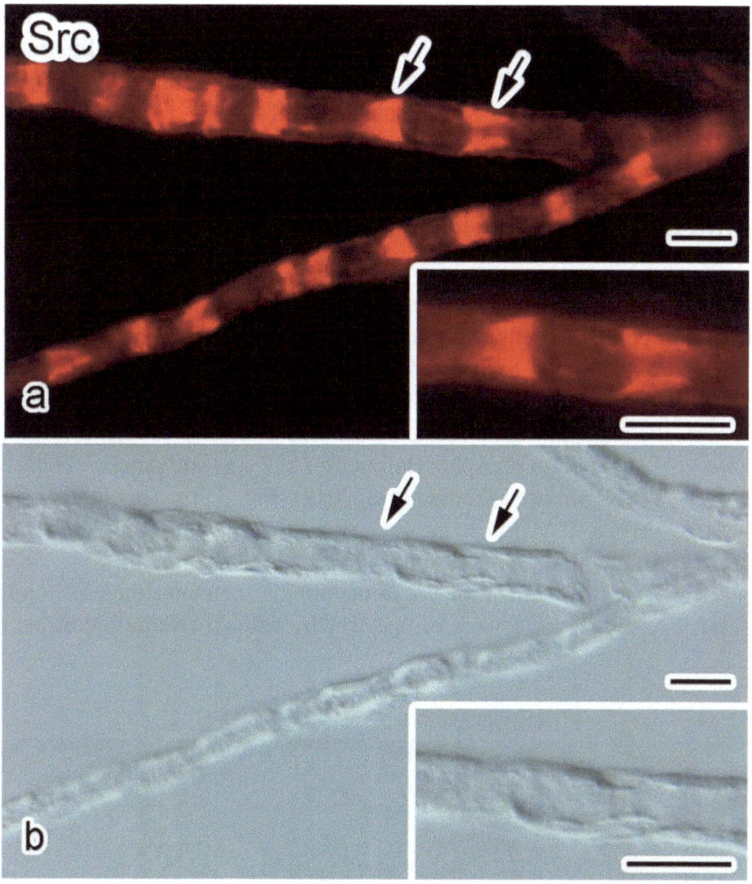

Fig. 3.6 Src immunostaining image in teased sciatic nerve fibers of normal adult mice. Src is immunolocalized at SLIs (*arrows* in **a**). **b** Diffraction interference contrast image. Both *insets* are highly magnified views of SLIs. *Bars* 10 μm

demonstrated that Src was localized in SLI. In addition, Src was phosphorylated in Y527, but not in Y416 (Terada et al. 2013) under normal conditions (Fig. 3.6).

Concerning the Src kinase family, the requirement of Src activity for junctional assembly was documented in Fyn–/– or Src/Fyn-deficient keratinocytes (Calautti et al. 1998). One possible role of Src is for adhesive contacts in relation to E-cadherin by activation and recruitment of phosphatidylinositol-3-kinase (PI3K) (Pang et al. 2005). A contrasting role for the requirement of Src kinase activity for junction disassembly was also suggested, following observation that expression of a dominant-negative Src promoted the stability of cadherin-dependent cell–cell contacts (Alanne et al. 2009). These biphasic demonstrations of the role of Src are thought to be an effect of molecular conformational changes from phosphorylation/dephosphorylation (McLachlan et al. 2007; McLachlan and Yap 2011). In particular, combination of phosphorylation of Y416 and dephosphorylation of Y527 has been well documented to stimulate complete activation of Src, which provides a binding site for the Src homology 2 (SH2) domain of other cellular proteins (Tatosyan and Mizenina 2000). Thus, it is interesting to examine molecular switching of the Src phosphorylation within Schwann cells.

3.8 Interaction of Src with MPP6 in Schwann Cells

The MAGUK protein MPP2 mentioned in Sect. 3.2 was reported to negatively regulate the Src in epithelial cells (Baumgartner et al. 2009). In Schwann cells, MPP6 was found to interact with Src (Terada et al. 2013). As demonstrated in Sect. 3.3, because the MPP6 localization was disrupted in the SLIs of the 4.1G−/− peripheral nerve fibers, the phosphorylation state of Src was examined in them. Interestingly, in 4.1G−/− SLIs, although phosphorylation of Y527 was still detected, phosphorylation of Y418 was increased (Terada et al. 2013) (Fig. 3.7b). This finding suggests that the 4.1G–MPP6–CADM4 membrane skeletal molecular complex has a functional relationship with the Src phosphorylation in SLI (Fig. 3.7c).

Concerning the involvement of Src in membrane skeletons, there is an interesting report; Src reduced cleavage of spectrin by calpain (Nedrelow et al. 2003).

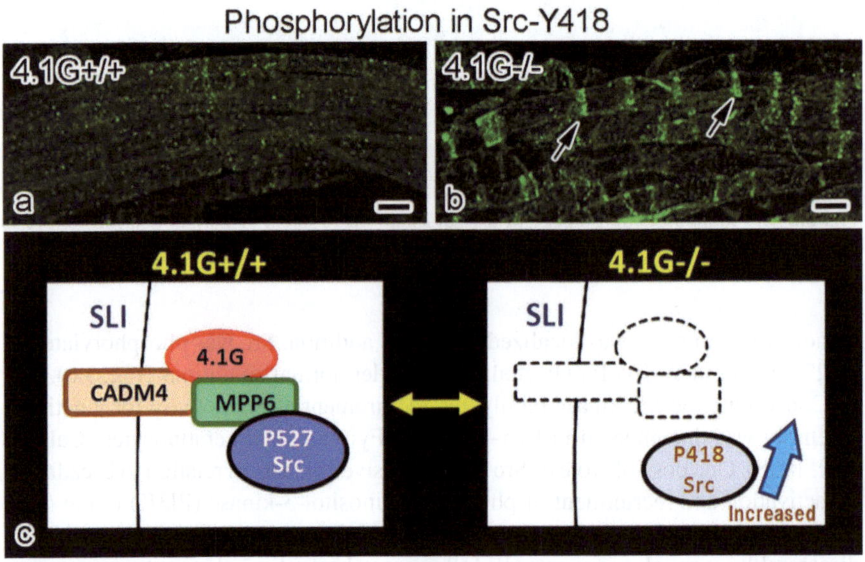

Fig. 3.7 Immunostaining images of phosphorylation in Src-Y418 in 4.1G+/+ (**a**, wild type) and 4.1G−/− (**b**) mouse sciatic nerve fibers, and schematic models of the relationship between the 4.1G–MPP6–CADM4 molecular complex and Src (**c**). The phosphorylation of Y418-Src appears more clearly in the SLIs of the 4.1G−/− nerve fibers (*arrows* in **b**). **c** Src in SLI is mainly phosphorylated at Y527 (P527 Src) in the normal 4.1G+/+ (wild-type) mouse, but the other Src is phosphorylated at Y418 (P418 Src) in the 4.1G−/− mouse. *Bars* 10 μm

Originally, 4.1R in erythrocytes was well documented to have an important role in resistance against external mechanical forces under blood flow (Gauthier et al. 2011). Such a role in membrane stability with a 4.1 family protein, 4.1B, was also demonstrated in axons of myelinated nerve fibers (Buttermore et al. 2011). Thus, combining the two facts about the MPP6–Src interaction and localization of Src, the Y527-phosphorylated Src–MPP6 complex has a potential role in the maintenance of membrane stability in SLIs.

3.9 Involvement of Src with Other Adhesion Molecules and Membrane Skeletal Components in Schwann Cells

Although Src in SLI interacts with 4.1G–MPP6–CADM4, it was still localized in SLIs of the 4.1G−/− nerve fibers (Terada et al. 2013). Combined with the finding that MPP6 did not localize in SLIs of 4.1G−/− nerve fibers, as mentioned in Sect. 3.3 (Ivanovic et al. 2012), this indicates that 4.1G–MPP6–CADM4 is not the only molecular complex interacting with Src in SLIs.

E-cadherin has been well documented to be influenced by p120 catenin (p120ctn) in SLI (Tricaud et al. 2005). In epithelial cells, Src was reported to regulate the E-cadherin–p120ctn molecular complex in epithelial cells (Reynolds 2007; Reynolds et al. 1992). One already proposed mechanism of Src is the E-cadherin internalization from the cell membrane to cytoplasm, as indicated under shear stress in metastatic esophageal cancer cells (Lawler et al. 2009). Another E-cadherin interacting protein, beta-catenin, is also known to be phosphorylated by Src, resulting in reduction in cell–cell adhesion in epithelial cells (Lilien and Balsamo 2005). In SLIs, the E-cadherin–p120ctn complex was proven to be an essential molecule revealed by experiments with mutant E-cadherin on the regulating region of p120ctn (Tricaud et al. 2005). However, E-cadherin deficiency in vivo seemed to show redundancy in the nerve fibers (Young et al. 2002). Therefore, we assume that Src may have a role in balancing the 4.1G–MPP6–CADM4 molecular complex and another E-cadherin–p120ctn molecular complex for stronger membrane–membrane adhesion in SLIs.

3.10 Phosphorylation State of Src in Schwanomma

Considering schwannoma, a kind of Schwann cell tumor, the abundant appearance of Src phosphorylation at Y418 was reported in *neurofibromin (Nf)*-2−/− cells that lost merlin (Houshmandi et al. 2009). In another melanoma cell line, RT4-D6P2T, this phenomenon was similarly observed (Terada et al. 2013), but Src phosphorylation of Y527 was not detected. Because such tumor cells have strong potential for migration and protrusion of their cell processes, we think that opposing phosphorylation states between in vivo SLI and schwanomma are probably related to the cell–cell (membrane–membrane in the case of SLI) or cell–matrix adhesion systems.

3.11 Involvement of Src in Schwann Cell SLIs Under Development and Wallerian Degeneration

Our findings about Src localization in SLI under normal conditions fill a gap in a previous report that shows abundant expression of active-Src in the cytoplasm of Schwann cells under Wallerian degeneration (Zhao et al. 2003). It is thought that SLI is a sensitive region under Wallerian degeneration (Jung et al. 2011). During Wallerian degeneration, a signaling molecule, Ras-related C3 botulimum toxin substrate 1 (Rac1), was demonstrated to induce F-actin reorganization (Park and Feltri 2011). Because Src is known to affect Rac1 (Chaturvedi et al. 2007), the Src-Rac1 molecular complex may be one of the initial warning signals for membrane stability in Schwann cells through SLI. Thus, we propose that detection of the Src phosphorylation state in SLI could be a good indicator to determine the maturity or health of myelinated nerve fibers.

For myelination in Schwann cells, neural Wiskott–Aldrich syndrome protein (N-WASP) was shown to regulate actin nucleation (Novak et al. 2011), and Src was documented as an activator of N-WASP phosphorylation (Suetsugu et al. 2002). By treatment with a Src inhibitor, 4-amino-5-(4-chlorophenyl)-7-(*t*-butyl) pyrazolo [3,4-*d*] pyrimidine (PP2), dose-dependent reduction in the accumulation of several myelin proteins, including myelin basic protein (MBP), myelin glycoprotein P_0, and myelin-associated glycoprotein (MAG), was demonstrated in rat Schwann cell–dorsal root ganglion neuron co-cultures (Hossain et al. 2010), which implies an effect of Src on the initiation and development of myelination.

Concerning the phosphorylation state of Src during development of Schwann cells, we reported the appearance of Src phosphorylation of Y418 in SLI before the Src phosphorylation of Y527 in postnatal mouse sciatic nerves (Terada et al. 2013). To prove our proposed concept that "the Src phosphorylation state in SLI could be an indicator to determine the maturity and/or health of myelinated nerve fibers," it would be interesting to examine Src function in relationship to SLI under some pathological changes, such as Wallerian degeneration and subsequent regeneration.

3.12 Concluding Remarks

The molecular complex defines cell and tissue structure, and some molecules have roles in signal transduction themselves. In this review, the membrane skeletal protein complex 4.1G–MPP6–CADM4 and its relationship to a signal molecule Src were demonstrated in SLI. We believe that examination of the relationship of this complex with other molecular complexes will reveal the morphofunctional mechanisms of SLIs in Schwann cells and bring a new perspective regarding pathological conditions.

References

Alanne MH, Pummi K, Heape AM, Grenman R, Peltonen J, Peltonen S (2009) Tight junction proteins in human Schwann cell autotypic junctions. J Histochem Cytochem 57:523–529

Baines AJ (2010) The spectrin-ankyrin-4.1-adducin membrane skeleton: adapting eukaryotic cells to the demands of animal life. Protoplasma 244:99–131

Baumgartner M, Weiss A, Fritzius T, Heinrich J, Moelling K (2009) The PDZ protein MPP2 interacts with c-Src in epithelial cells. Exp Cell Res 315:2888–2898

Bensenor LB, Barlan K, Rice SE, Fehon RG, Gelfand VI (2010) Microtubule-mediated transport of the tumor-suppressor protein Merlin and its mutants. Proc Natl Acad Sci USA 107: 7311–7316

Biederer T, Sudhof TC (2001) CASK and protein 4.1 support F-actin nucleation on neurexins. J Biol Chem 276:47869–47876

Bohl J, Brimer N, Lyons C, Vande Pol SB (2007) The stardust family protein MPP7 forms a tripartite complex with LIN7 and DLG1 that regulates the stability and localization of DLG1 to cell junctions. J Biol Chem 282:9392–9400

Bolino A, Bolis A, Previtali SC, Dina G, Bussini S, Dati G, Amadio S, Del Carro U, Mruk DD, Feltri ML, Cheng CY, Quattrini A, Wrabetz L (2004) Disruption of Mtmr2 produces CMT4B1-like neuropathy with myelin outfolding and impaired spermatogenesis. J Cell Biol 167: 711–721

Bolis A, Coviello S, Visigalli I, Taveggia C, Bachi A, Chishti AH, Hanada T, Quattrini A, Previtali SC, Biffi A, Bolino A (2009) Dlg1, Sec8, and Mtmr2 regulate membrane homeostasis in Schwann cell myelination. J Neurosci 29:8858–8870

Buttermore ED, Dupree JL, Cheng J, An X, Tessarollo L, Bhat MA (2011) The cytoskeletal adaptor protein band 4.1B is required for the maintenance of paranodal axoglial septate junctions in myelinated axons. J Neurosci 31:8013–8024

Calautti E, Cabodi S, Stein PL, Hatzfeld M, Kedersha N, Paolo Dotto G (1998) Tyrosine phosphorylation and src family kinases control keratinocyte cell-cell adhesion. J Cell Biol 141: 1449–1465

Chaturvedi LS, Marsh HM, Shang X, Zheng Y, Basson MD (2007) Repetitive deformation activates focal adhesion kinase and ERK mitogenic signals in human Caco-2 intestinal epithelial cells through Src and Rac1. J Biol Chem 282:14–28

Chen J, Terada N, Ohno N, Saitoh S, Saitoh Y, Ohno S (2011) Immunolocalization of membrane skeletal protein, 4.1G, in enteric glial cells in the mouse large intestine. Neurosci Lett 488: 193–198

Cotter L, Ozcelik M, Jacob C, Pereira JA, Locher V, Baumann R, Relvas JB, Suter U, Tricaud N (2010) Dlg1-PTEN interaction regulates myelin thickness to prevent damaging peripheral nerve overmyelination. Science 328:1415–1418

Da Costa L, Galimand J, Fenneteau O, Mohandas N (2013) Hereditary spherocytosis, elliptocytosis, and other red cell membrane disorders. Blood Rev 27(4):167–178

de Mendoza A, Suga H, Ruiz-Trillo I (2010) Evolution of the MAGUK protein gene family in premetazoan lineages. BMC Evol Biol 10:93

Discher DE, Winardi R, Schischmanoff PO, Parra M, Conboy JG, Mohandas N (1995) Mechanochemistry of protein 4.1's spectrin-actin-binding domain: ternary complex interactions, membrane binding, network integration, structural strengthening. J Cell Biol 130: 897–907

Fujita E, Kouroku Y, Ozeki S, Tanabe Y, Toyama Y, Maekawa M, Kojima N, Senoo H, Toshimori K, Momoi T (2006) Oligo-astheno-teratozoospermia in mice lacking RA175/TSLC1/SynCAM/IGSF4A, a cell adhesion molecule in the immunoglobulin superfamily. Mol Cell Biol 26:718–726

Funke L, Dakoji S, Bredt DS (2005) Membrane-associated guanylate kinases regulate adhesion and plasticity at cell junctions. Annu Rev Biochem 74:219–245

Gauthier E, Guo X, Mohandas N, An X (2011) Phosphorylation-dependent perturbations of the 4.1R-associated multiprotein complex of the erythrocyte membrane. Biochemistry 50:4561–4567

Goebbels S, Oltrogge JH, Kemper R, Heilmann I, Bormuth I, Wolfer S, Wichert SP, Mobius W, Liu X, Lappe-Siefke C, Rossner MJ, Groszer M, Suter U, Frahm J, Boretius S, Nave KA (2010) Elevated phosphatidylinositol 3,4,5-trisphosphate in glia triggers cell-autonomous membrane wrapping and myelination. J Neurosci 30:8953–8964

Hossain S, Fragoso G, Mushynski WE, Almazan G (2010) Regulation of peripheral myelination by Src-like kinases. Exp Neurol 226:47–57

Houshmandi SS, Emnett RJ, Giovannini M, Gutmann DH (2009) The neurofibromatosis 2 protein, merlin, regulates glial cell growth in an ErbB2- and Src-dependent manner. Mol Cell Biol 29:1472–1486

Ivanovic A, Horresh I, Golan N, Spiegel I, Sabanay H, Frechter S, Ohno S, Terada N, Mobius W, Rosenbluth J, Brose N, Peles E (2012) The cytoskeletal adapter protein 4.1G organizes the internodes in peripheral myelinated nerves. J Cell Biol 196:337–344

Jung J, Cai W, Lee HK, Pellegatta M, Shin YK, Jang SY, Suh DJ, Wrabetz L, Feltri ML, Park HT (2011) Actin polymerization is essential for myelin sheath fragmentation during Wallerian degeneration. J Neurosci 31:2009–2015

Kamberov E, Makarova O, Roh M, Liu A, Karnak D, Straight S, Margolis B (2000) Molecular cloning and characterization of Pals, proteins associated with mLin-7. J Biol Chem 275:11425–11431

Lawler K, O'Sullivan G, Long A, Kenny D (2009) Shear stress induces internalization of E-cadherin and invasiveness in metastatic oesophageal cancer cells by a Src-dependent pathway. Cancer Sci 100:1082–1087

Lilien J, Balsamo J (2005) The regulation of cadherin-mediated adhesion by tyrosine phosphorylation/dephosphorylation of beta-catenin. Curr Opin Cell Biol 17:459–465

Lozovatsky L, Abayasekara N, Piawah S, Walther Z (2009) CASK deletion in intestinal epithelia causes mislocalization of LIN7C and the DLG1/Scrib polarity complex without affecting cell polarity. Mol Biol Cell 20:4489–4499

Lue RA, Marfatia SM, Branton D, Chishti AH (1994) Cloning and characterization of hdlg: the human homologue of the Drosophila discs large tumor suppressor binds to protein 4.1. Proc Natl Acad Sci USA 91:9818–9822

Lue RA, Brandin E, Chan EP, Branton D (1996) Two independent domains of hDlg are sufficient for subcellular targeting: the PDZ1-2 conformational unit and an alternatively spliced domain. J Cell Biol 135:1125–1137

Machnicka B, Czogalla A, Hryniewicz-Jankowska A, Boguslawska DM, Grochowalska R, Heger E, Sikorski AF (2013) Spectrins: a structural platform for stabilization and activation of membrane channels, receptors and transporters. Biochim Biophys Acta, doi: 10.1016/j.bbamem.2013.05.002

Maurel P, Einheber S, Galinska J, Thaker P, Lam I, Rubin MB, Scherer SS, Murakami Y, Gutmann DH, Salzer JL (2007) Nectin-like proteins mediate axon Schwann cell interactions along the internode and are essential for myelination. J Cell Biol 178:861–874

McLachlan RW, Yap AS (2011) Protein tyrosine phosphatase activity is necessary for E-cadherin-activated Src signaling. Cytoskeleton 68:32–43

McLachlan RW, Kraemer A, Helwani FM, Kovacs EM, Yap AS (2007) E-cadherin adhesion activates c-Src signaling at cell-cell contacts. Mol Biol Cell 18:3214–3223

Mohandas N, Gallagher PG (2008) Red cell membrane: past, present, and future. Blood 112:3939–3948

Nedrelow JH, Cianci CD, Morrow JS (2003) c-Src binds alpha II spectrin's Src homology 3 (SH3) domain and blocks calpain susceptibility by phosphorylating Tyr1176. J Biol Chem 278:7735–7741

Nix SL, Chishti AH, Anderson JM, Walther Z (2000) hCASK and hDlg associate in epithelia, and their src homology 3 and guanylate kinase domains participate in both intramolecular and intermolecular interactions. J Biol Chem 275:41192–41200

Novak N, Bar V, Sabanay H, Frechter S, Jaegle M, Snapper SB, Meijer D, Peles E (2011) N-WASP is required for membrane wrapping and myelination by Schwann cells. J Cell Biol 192:243–250

Nunomura W, Takakuwa Y, Parra M, Conboy J, Mohandas N (2000) Regulation of protein 4.1R, p55, and glycophorin C ternary complex in human erythrocyte membrane. J Biol Chem 275:24540–24546

Ogita H, Takai Y (2006) Nectins and nectin-like molecules: roles in cell adhesion, polarization, movement, and proliferation. IUBMB Life 58:334–343

Ohno S, Terada N, Fujii Y, Ueda H (1994) Membrane skeleton in fresh unfixed erythrocytes as revealed by a rapid-freezing and deep-etching method. J Anat 185:415–420

Ohno S, Terada N, Fujii Y, Ueda H, Takayama I (1996) Dynamic structure of glomerular capillary loop as revealed by an in vivo cryotechnique. Virchows Arch 427:519–527

Ohno N, Terada N, Yamakawa H, Komada M, Ohara O, Trapp BD, Ohno S (2006) Expression of protein 4.1G in Schwann cells of the peripheral nervous system. J Neurosci Res 84:568–577

Ohno N, Terada N, Komada M, Saitoh S, Costantini F, Pace V, Germann PG, Weber K, Yamakawa H, Ohara O, Ohno S (2009) Dispensable role of protein 4.1B/DAL-1 in rodent adrenal medulla regarding generation of pheochromocytoma and plasmalemmal localization of TSLC1. Biochim Biophys Acta 1793:506–515

Ohno S, Terada N, Ohno N, Saitoh S, Saitoh Y, Fujii Y (2010) Significance of 'in vivo cryotechnique' for morphofunctional analyses of living animal organs. J Electron Microsc 59:395–408

Ozcelik M, Cotter L, Jacob C, Pereira JA, Relvas JB, Suter U, Tricaud N (2010) Pals1 is a major regulator of the epithelial-like polarization and the extension of the myelin sheath in peripheral nerves. J Neurosci 30:4120–4131

Pang JH, Kraemer A, Stehbens SJ, Frame MC, Yap AS (2005) Recruitment of phosphoinositide 3-kinase defines a positive contribution of tyrosine kinase signaling to E-cadherin function. J Biol Chem 280:3043–3050

Park HT, Feltri ML (2011) Rac1 GTPase controls myelination and demyelination. Bioarchitecture 1:110–113

Parra M, Gascard P, Walensky LD, Snyder SH, Mohandas N, Conboy JG (1998) Cloning and characterization of 4.1G (EPB41L2), a new member of the skeletal protein 4.1 (EPB41) gene family. Genomics 49:298–306

Peters LL, Weier HU, Walensky LD, Snyder SH, Parra M, Mohandas N, Conboy JG (1998) Four paralogous protein 4.1 genes map to distinct chromosomes in mouse and human. Genomics 54:348–350

Poliak S, Matlis S, Ullmer C, Scherer SS, Peles E (2002) Distinct claudins and associated PDZ proteins form different autotypic tight junctions in myelinating Schwann cells. J Cell Biol 159:361–372

Quinn BJ, Welch EJ, Kim AC, Lokuta MA, Huttenlocher A, Khan AA, Kuchay SM, Chishti AH (2009) Erythrocyte scaffolding protein p55/MPP1 functions as an essential regulator of neutrophil polarity. Proc Natl Acad Sci USA 106:19842–19847

Reynolds AB (2007) p120-catenin: past and present. Biochim Biophys Acta 1773:2–7

Reynolds AB, Herbert L, Cleveland JL, Berg ST, Gaut JR (1992) p120, a novel substrate of protein tyrosine kinase receptors and of p60v-src, is related to cadherin-binding factors beta-catenin, plakoglobin and armadillo. Oncogene 7:2439–2445

Saitoh Y, Terada N, Saitoh S, Ohno N, Jin T, Ohno S (2012) Histochemical analyses and quantum dot imaging of microvascular blood flow with pulmonary edema in living mouse lungs by "in vivo cryotechnique". Histochem Cell Biol 137:137–151

Scheiermann C, Meda P, Aurrand-Lions M, Madani R, Yiangou Y, Coffey P, Salt TE, Ducrest-Gay D, Caille D, Howell O, Reynolds R, Lobrinus A, Adams RH, Yu AS, Anand P, Imhof BA, Nourshargh S (2007) Expression and function of junctional adhesion molecule-C in myelinated peripheral nerves. Science 318:1472–1475

Scherer SS, Gutmann DH (1996) Expression of the neurofibromatosis 2 tumor suppressor gene product, merlin, in Schwann cells. J Neurosci Res 46:595–605
Seo PS, Jeong JJ, Zeng L, Takoudis CG, Quinn BJ, Khan AA, Hanada T, Chishti AH (2009) Alternatively spliced exon 5 of the FERM domain of protein 4.1R encodes a novel binding site for erythrocyte p55 and is critical for membrane targeting in epithelial cells. Biochim Biophys Acta 1793:281–289
Shindo M, Wada H, Kaido M, Tateno M, Aigaki T, Tsuda L, Hayashi S (2008) Dual function of Src in the maintenance of adherens junctions during tracheal epithelial morphogenesis. Development (Camb) 135:1355–1364
Shingai T, Ikeda W, Kakunaga S, Morimoto K, Takekuni K, Itoh S, Satoh K, Takeuchi M, Imai T, Monden M, Takai Y (2003) Implications of nectin-like molecule-2/IGSF4/RA175/SgIGSF/TSLC1/SynCAM1 in cell–cell adhesion and transmembrane protein localization in epithelial cells. J Biol Chem 278:35421–35427
Spiegel I, Adamsky K, Eshed Y, Milo R, Sabanay H, Sarig-Nadir O, Horresh I, Scherer SS, Rasband MN, Peles E (2007) A central role for Necl4 (SynCAM4) in Schwann cell-axon interaction and myelination. Nat Neurosci 10:861–869
Suetsugu S, Hattori M, Miki H, Tezuka T, Yamamoto T, Mikoshiba K, Takenawa T (2002) Sustained activation of N-WASP through phosphorylation is essential for neurite extension. Dev Cell 3:645–658
Surace EI, Strickland A, Hess RA, Gutmann DH, Naughton CK (2006) Tslc1 (nectin-like molecule-2) is essential for spermatozoa motility and male fertility. J Androl 27:816–825
Susuki K, Raphael AR, Ogawa Y, Stankewich MC, Peles E, Talbot WS, Rasband MN (2011) Schwann cell spectrins modulate peripheral nerve myelination. Proc Natl Acad Sci USA 108:8009–8014
Tatosyan AG, Mizenina OA (2000) Kinases of the Src family: structure and functions. Biochemistry (Mosc) 65:49–58
Terada N, Fujii Y, Ueda H, Ohno S (1997) An immunocytochemical study of changes in the human erythrocyte membrane skeleton produced by stretching examined by the quick-freezing and deep-etching method. J Anat 190:397–404
Terada N, Fujii Y, Kato Y, Ueda H, Baba T, Ohno S (1998a) Scanning electron microscopic study of erythrocyte shapes artificially jetted through tubes at different pressures by 'in vitro cryotechnique for erythrocytes'. J Electron Microsc 47:489–493
Terada N, Kato Y, Fuji Y, Ueda H, Baba T, Ohno S (1998b) Scanning electron microscopic study of flowing erythrocytes in hepatic sinusoids as revealed by 'in vivo cryotechnique'. J Electron Microsc 47:67–72
Terada N, Ohno N, Saitoh S, Saitoh Y, Komada M, Kubota H, Ohno S (2010) Involvement of a membrane skeletal protein, 4.1G, for Sertoli/germ cell interaction. Reproduction 139:883–892
Terada N, Saitoh Y, Ohno N, Komada M, Saitoh S, Peles E, Ohno S (2012a) Essential function of protein 4.1G in targeting of membrane protein palmitoylated 6 into Schmidt-Lanterman incisures in myelinated nerves. Mol Cell Biol 32:199–205
Terada N, Saitoh Y, Saitoh S, Ohno N, Fujishita K, Koizumi S, Ohno S (2012b) Visualization of ATP with luciferin-luciferase reaction in mouse skeletal muscles using an "in vivo cryotechnique". Microsc Microanal 18:1030–1036
Terada N, Saitoh Y, Ohno N, Komada M, Yamauchi J, Ohno S (2013) Involvement of Src in the membrane skeletal complex, MPP6-4.1G, in Schmidt-Lanterman incisures of mouse myelinated nerve fibers in PNS. Histochem Cell Biol 140:213–222
Trapp BD, Andrews SB, Wong A, O'Connell M, Griffin JW (1989) Co-localization of the myelin-associated glycoprotein and the microfilament components, F-actin and spectrin, in Schwann cells of myelinated nerve fibres. J Neurocytol 18:47–60
Tricaud N, Perrin-Tricaud C, Bruses JL, Rutishauser U (2005) Adherens junctions in myelinating Schwann cells stabilize Schmidt-Lanterman incisures via recruitment of p120 catenin to E-cadherin. J Neurosci 25:3259–3269

Tseng TC, Marfatia SM, Bryant PJ, Pack S, Zhuang Z, O'Brien JE, Lin L, Hanada T, Chishti AH (2001) VAM-1: a new member of the MAGUK family binds to human Veli-1 through a conserved domain. Biochim Biophys Acta 1518:249–259

van der Weyden L, Arends MJ, Chausiaux OE, Ellis PJ, Lange UC, Surani MA, Affara N, Murakami Y, Adams DJ, Bradley A (2006) Loss of TSLC1 causes male infertility due to a defect at the spermatid stage of spermatogenesis. Mol Cell Biol 26:3595–3609

Verhey KJ, Rapoport TA (2001) Kinesin carries the signal. Trends Biochem Sci 26:545–550

Wakayama T, Ohashi K, Mizuno K, Iseki S (2001) Cloning and characterization of a novel mouse immunoglobulin superfamily gene expressed in early spermatogenic cells. Mol Reprod Dev 60:158–164

Walensky LD, Gascard P, Fields ME, Blackshaw S, Conboy JG, Mohandas N, Snyder SH (1998) The 13-kD FK506 binding protein, FKBP13, interacts with a novel homologue of the erythrocyte membrane cytoskeletal protein 4.1. J Cell Biol 141:143–153

Wang Q, Chen XW, Margolis B (2007) PALS1 regulates E-cadherin trafficking in mammalian epithelial cells. Mol Biol Cell 18:874–885

Xu HM, Gutmann DH (1998) Merlin differentially associates with the microtubule and actin cytoskeleton. J Neurosci Res 51:403–415

Yamada D, Yoshida M, Williams YN, Fukami T, Kikuchi S, Masuda M, Maruyama T, Ohta T, Nakae D, Maekawa A, Kitamura T, Murakami Y (2006) Disruption of spermatogenic cell adhesion and male infertility in mice lacking TSLC1/IGSF4, an immunoglobulin superfamily cell adhesion molecule. Mol Cell Biol 26:3610–3624

Yamada KH, Hanada T, Chishti AH (2007) The effector domain of human Dlg tumor suppressor acts as a switch that relieves autoinhibition of kinesin-3 motor GAKIN/KIF13B. Biochemistry 46:10039–10045

Yang S, Guo X, Debnath G, Mohandas N, An X (2009) Protein 4.1R links E-cadherin/beta-catenin complex to the cytoskeleton through its direct interaction with beta-catenin and modulates adherens junction integrity. Biochim Biophys Acta 1788:1458–1465

Young P, Boussadia O, Berger P, Leone DP, Charnay P, Kemler R, Suter U (2002) E-cadherin is required for the correct formation of autotypic adherens junctions of the outer mesaxon but not for the integrity of myelinated fibers of peripheral nerves. Mol Cell Neurosci 21:341–351

Zhao YL, Takagawa K, Oya T, Yang HF, Gao ZY, Kawaguchi M, Ishii Y, Sasaoka T, Owada K, Furuta I, Sasahara M (2003) Active Src expression is induced after rat peripheral nerve injury. Glia 42:184–193

Chapter 4
Schwann Cell–Axon Interactions: The Molecular and Metabolic Link Between Schwann Cells and Axons

Nobuhiko Ohno, Takashi Sakoh, Yurika Saitoh, Nobuo Terada, and Shinichi Ohno

Abstract Schwann cells and axons have close and complex interactions that determine Schwann cell behavior and fate and support or impair axonal integrity. The interactions are mediated by molecules that are responsible for physical junctions between Schwann cells and axons and also soluble mediators which are generated and bidirectionally transported in the interface. Multiple types of axonal signals are critical for regulating Schwann cell proliferation, differentiation, myelination, and myelin maintenance. At the same time, Schwann cells regulate axonal development and play essential roles for survival of axons. Current evidence suggests that the trophic support of Schwann cells is associated with modulation of axonal metabolism, which is involved in functional maintenance of axonal mitochondria. Further advancement in genetic techniques, transgenic models, and myelinating cultures will elucidate the molecular and cellular mechanisms of Schwann cell–axon interactions that could lead to new therapies of peripheral nervous system diseases.

Keywords Axonal survival • Demyelination • Dysmyelination • Glycolysis • Lactate • Mitochondria • Myelination • Na^+/K^+-ATPase • Neuregulin • Trophic support

---

N. Ohno (✉) • T. Sakoh • Y. Saitoh • S. Ohno
Department of Anatomy and Molecular Histology, Interdisciplinary Graduate School of Medicine and Engineering, University of Yamanashi, Shimokato 1110, Chuo-City, Yamanashi 409-3898, Japan
e-mail: nohno@yamanashi.ac.jp

N. Terada
Department of Occupational Therapy, School of Health Sciences, Shinshu University School of Medicine, Matsumoto-City, Nagano, Japan

K. Sango and J. Yamauchi (eds.), *Schwann Cell Development and Pathology*,
DOI 10.1007/978-4-431-54764-8_4, © Springer Japan 2014

4.1 Introduction

Schwann cells, the glial cells in the peripheral nervous system (PNS), are among the largest and most ultrastructurally sophisticated cells in the body, and can undergo rapid and dynamic transformation during development as well as after injury. The complex structures and dynamic behavior of Schwann cells determine the way of interaction with axons, neuronal processes that confer nerve impulses to the target cells in the PNS (Hoke et al. 2006; Jessen and Mirsky 2005; Meyer et al. 1992; Mirsky et al. 2008; Webster Hde 1971). Initially in development, Schwann cells surround the external margins of the axonal bundles, and support axonal outgrowth by providing growth factors, such as nerve growth factor (NGF) and brain-derived neurotrophic factor (BDNF) (Fig. 4.1a). Thereafter, Schwann cells segregate the axons into successively smaller bundles. Finally, individual axons are covered by Schwann cell cytoplasm and separated from one another near the Schwann cell surface. Many PNS axons including small sensory and autonomic axons are kept in this state, and these units composed of small-diameter axons and their ensheathing Schwann cells are called unmyelinated or "Remak" fibers (Fig. 4.1b). On the other hand, other Schwann cells spirally wrap single axons, and form multilamellar membranes called myelin, which is an essential structure for rapid salutatory conduction (Fig. 4.1c, d).

Development of myelin appears later after evolution of primitive members of the vertebrate line, whereas glial ensheathment around axons itself is an early feature of nervous system evolution (Hartline and Colman 2007). Indeed, glial cells in the invertebrate nervous system engulf multiple axons without myelinating them and appear similar to the non-myelin-forming Schwann cells in vertebrates (Klambt et al. 2001). By contrast, the PNS and central nervous system (CNS) of vertebrates have many myelinated axons with clustering of membrane-associated proteins including

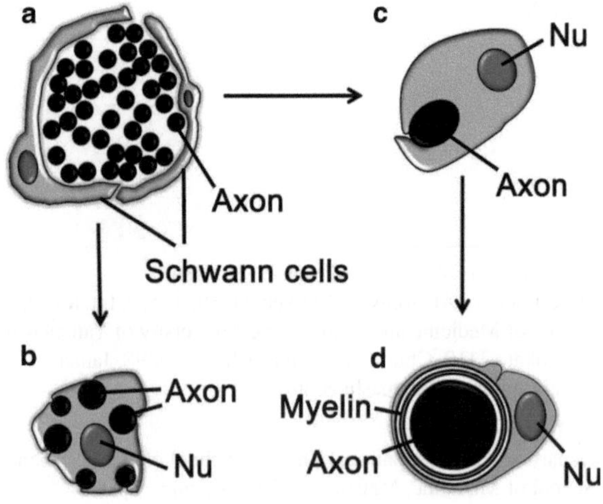

Fig. 4.1 Schematic drawing of developmental changes in Schwann cell morphology. Immature Schwann cells engulf the bulk of axons (**a**). Schwann cells ensheath small-diameter axons and differentiate to make unmyelinated Remak bundles (**b**), whereas those ensheathing the large-diameter axons produce myelin (**c**, **d**). *Nu* Schwann cell nuclei

adhesion molecules and ion channels in differential axonal segments divided by distinct myelin domains (Salzer 2003). A large part of Schwann cell development and fate determination in the vertebral nervous system including myelin formation is controlled by interactions between Schwann cells and axons. For example, the initiation of myelin formation by Schwann cells is strictly dependent on axonal signals, in contrast to myelin-forming cells in the CNS, the oligodendrocytes (Birchmeier and Nave 2008). At the same time, recent studies elucidated the metabolic relationship between Schwann cells and axons, which regulates the axonal microenvironment and affects axonal survival (Nave 2010b). Interestingly, certain aspects of glial support for axonal integrity and survival appear to be common in PNS and CNS, indicating that different sets of glial cells have similar roles for ensheathed axons. The principal goal of this chapter is to provide an overview of recent findings regarding the molecular and metabolic link between Schwann cells and axons, which significantly affect Schwann cell behavior as well as axonal integrity.

4.2 Structural Interactions Between Schwann Cells and Axons

Myelination of Schwann cells is extreme cellular specialization of the vertebrate nervous system. Myelination requires the generation of large amounts of extended cell membranes to ensheath around axons many times. The compacted and insulating sheath of myelin leaves gaps for the highly specialized nodes of Ranvier (Fig. 4.2a). At the nodes, voltage-dependent Na^+ channels are clustered on the axolemma, and focal depolarization of these Na^+ channels is responsible for saltatory conduction. During development, heminodes are first formed at the longitudinally expanding edges of single Schwann cells. Na^+ channels and nodal proteins are localized in the adjacent regions of the outermost edges of Schwann cells. Thereafter, the nodal components are moved ahead together with the edges of Schwann cells, and finally form mature nodes with nodal components of adjacent Schwann cells. Nodal regions are formed and maintained by scaffolding and adhesion molecules. For example, interaction between a membrane-bound extracellular matrix, gliomedin, produced from Schwann cells and axolemmal cell adhesion molecules, including neurofascin 186 (NF186) and NrCAM, is an early event in nodal formation (Salzer et al. 2008). When these molecules are genetically disrupted, the formation of the nodes such as Na^+-channel clustering is significantly impaired or delayed (Eshed et al. 2007; Feinberg et al. 2010; Sherman et al. 2005).

In the PNS, axonal regions between two adjacent nodes are covered with myelin sheath and composed of different types of segments with specialized functions. The nodes of Ranvier are flanked by paranodes that have axo–glial junctions which limit the diffusion of small molecules (Fig. 4.2b) (Perkins et al. 2008; Rosenbluth 2009). In paranodes, Schwann cell membranes are closely juxtaposed to axolemma and form a “paranodal loop”. Paranodal loops and axolemma are separated by a distance of 2.5–3.0 nm and connected by high electron densities that represent septate-like

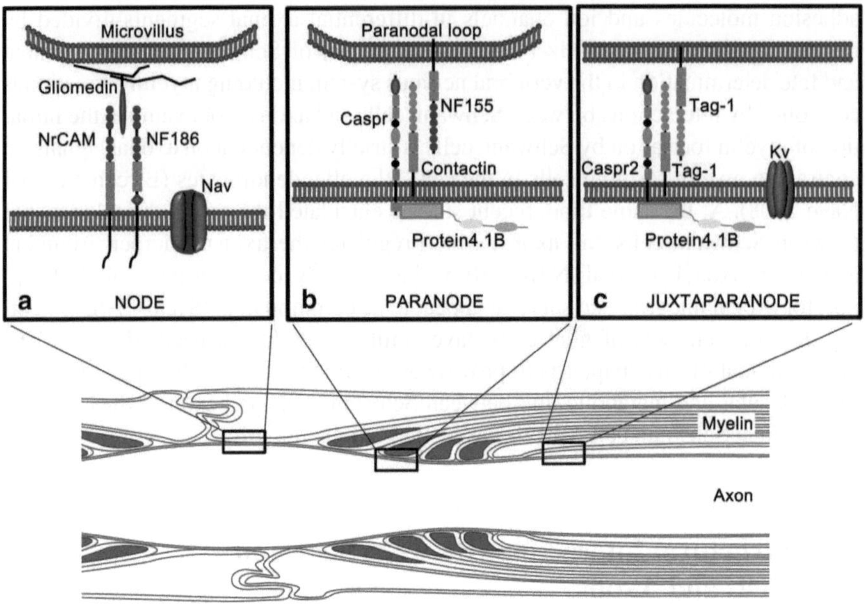

Fig. 4.2 Schematic drawing shows junctional complexes between Schwann cells and axons in different segments of myelinated axons. Nodal axolemma with voltage-gated Na^+ channels (*Nav*) has NrCAM and NF186, which interact with gliomedin (**a**). Paranodal axolemma contains Caspr and contactin, which interact with NF155 on paranodal loops of Schwann cells and are tethered to axonal actin cytoskeletons via protein 4.1B (**b**). Juxtaparanodal axolemma with enriched K^+ channels is associated with Schwann cell membranes via Tag-1 and Caspr2, which are connected to axonal actin cytoskeletons via protein 4.1B (**c**)

junctions (Peters et al. 1991). The septate-like junctions are the adhesive apparatus between myelin and axons, composed of contactin and contactin-associated protein (Caspr), the axonal membrane proteins enriched in paranodal regions (Einheber et al. 1997; Menegoz et al. 1997; Rios et al. 2000). These contactin–Caspr complexes are tethered to membrane-associated actin cytoskeletons via protein 4.1B (Buttermore et al. 2011; Denisenko-Nehrbass et al. 2003). Paranodal membranes of Schwann cells contain neurofascin 155 (NF155), which interacts with contactin and Caspr (Charles et al. 2002; Tait et al. 2000). The critical roles of paranodal septate-like junctions have been well documented through analyses of mice lacking components of the septate-like junctions (Bhat et al. 2001; Boyle et al. 2001; Sherman et al. 2005). Paranodal septate-like junctions are absent or abnormal in mice lacking Caspr and contactin. Nerve conduction in those mice is significantly slowed with accumulation of intracellular organelles in nodal/paranodal axoplasm. Although clustering of nodal Na^+ channels appears to be unchanged in contactin-null mice, voltage-gated K^+ channels normally excluded from paranodal axolemma in wild-type mice are diffused into paranodal axolemma in Caspr- or contactin-knockout mice (Bhat et al. 2001; Boyle et al. 2001). The paranodal axo–glial junction therefore functions as a "diffusion barrier" and spatially separates Na^+ and K^+ channel distribution in myelinated axons. Extracellularly, the spiral loops of paranodal segments provide a diffusion pathway to small soluble molecules, and their diffusion does not appear to be affected by lack of the paranodal junctions (Mackenzie et al. 1984; Mierzwa et al. 2011; Shroff et al. 2011).

The juxtaparanode is adjacent to the paranodal region and contains enriched voltage-gated fast K^+ channels. Although the adaxonal membranes of Schwann cells in internodes are usually smooth, juxtaparanodal membranes can form invaginations into the axons. In a variety of neuropathies, the invaginations of Schwann cell processes into the axoplasm become extensive (Griffin and Price 1981; Spencer and Thomas 1974). Tag-1 (transient axonal glycoprotein-1/contactin-2) on the juxtaparanodal membranes of Schwann cells and axons is considered to mediate Schwann cell–axon connection in the juxtaparanode and also to form complexes with Caspr2 (Fig. 4.2c) (Poliak et al. 1999; Traka et al. 2003). The Caspr2–Tag-1 complex is tethered to membrane-associated actin cytoskeletons via protein 4.1B (Denisenko-Nehrbass et al. 2003). These interactions in addition to paranodal septate-like junctions limit the diffusion of K^+ channels to the nodal region (Bhat et al. 2001; Traka et al. 2003).

Membrane-associated proteins are also localized at the interface between Schwann cells and axons and influence Schwann cell differentiation. Necl-4 (nectin-like protein-4) on Schwann cells binds to Necl-1 on axons and is considered to facilitate myelination, although mice lacking Necl-1 have little deficit in PNS myelination (Maurel et al. 2007; Park et al. 2008; Spiegel et al. 2007). Localization of Par3 and its interaction with p75NTR (neurotrophin receptor) at the axon–Schwann cell junction are crucial to start myelination during development, suggesting that neurotrophins play some modulatory roles in myelination of Schwann cells (Chan et al. 2006; Xiao et al. 2009). A cell adhesion molecule of Schwann cells, N-cadherin, colocalizes with Par3 at the axon–Schwann cell interface upon myelination and may mediate the recruitment of Par3 to the interface, as seen in epithelial cells (Lewallen et al. 2011). Because myelination is delayed in mice with Schwann cell-specific depletion of N-cadherin and its associated molecule, β-catenin, N-cadherin along with β-catenin may be involved in establishment of Schwann cell polarity and the timing of myelination. However, there are some redundant factors for both proteins in the formation and maturation of myelin (Lewallen et al. 2011).

Nonmyelinating Schwann cells forming Remak bundles lack myelin and myelin components but express cell adhesion molecules and cell-surface receptors that are less abundant in myelinating cells (Mirsky et al. 2008). The cell adhesion molecules, L1 and N-CAM, are abundant in Remak Schwann cells but are downregulated upon myelination. L1 expression by Schwann cells is essential for Schwann cell contact and survival of sensory axons (Haney et al. 1999). N-CAM is a 120- to 180-kDa glycoprotein that is related to axonal outgrowth (Martini 1994). The interaction between nonmyelinating Schwann cells and axons may also have specific functions to maintain and modulate the periaxonal ionic microenvironment, such as K^+ regulation (Robert and Jirounek 1994).

Schwann cells produce extracellular matrix molecules, such as collagens, laminins, fibronectin, and heparan-sulfated proteoglycans (Carey and Todd 1987;

Chernousov et al. 2008; Feltri and Wrabetz 2005; Rasi et al. 2010). Such synthesis of extracellular matrix components by Schwann cells organizes basal lamina as a sleeve around the Schwann cells. Interaction with axons and production of the basal lamina is interdependent, because axons are important to facilitate secretion of basal lamina components by Schwann cells, and the basal lamina is required for efficient differentiation of Schwann cells and myelination (Bunge et al. 1986; Carey et al. 1983; Eldridge et al. 1987; Fernandez-Valle et al. 1993; Podratz et al. 2001). Integrins functionally mediate interaction of Schwann cells with extracellular matrix, and impaired integrin function or ligand production perturbs their axonal radial sorting and myelination (Chernousov et al. 2008; Rasi et al. 2010). The basal lamina of Schwann cells is also an important factor affecting nerve regeneration in the PNS after injury, as its components such as laminin are substrates supporting axonal outgrowth and guidance (Bunge et al. 1989).

4.3 Axonal Signaling for Modulation of Schwann Cell Behavior

In the PNS, signals from axons control the development of Schwann cells, including proliferation of precursors derived from neural crest and differentiation for myelin formation (Jessen and Mirsky 2005). Axonal neuregulins (NRG) are a family of cell signaling molecules that regulate proliferation, differentiation, and survival of Schwann cells and interact with receptor tyrosine kinase receptor ErbB (Dong et al. 1995; Grinspan et al. 1996; Morrissey et al. 1995; Nave and Trapp 2008; Newbern and Birchmeier 2010; Trachtenberg and Thompson 1996). Deficiency of NRG isoforms in PNS or ErbB receptor complexes in Schwann cells decreases the number of Schwann cells (Morris et al. 1999; Riethmacher et al. 1997; Woldeyesus et al. 1999). NRG1 has at least 15 isoforms, and membrane-bound type III isoforms of NRG1 appear to be key regulaters of axon–Schwann cell signaling for myelination. NRG1 type III binds to ErbB2–ErbB3 receptor complexes in Schwann cells and determines the threshold triggering myelination and myelin thickness matching to axon caliber (Birchmeier and Nave 2008; Carroll et al. 1997; Cohen et al. 1992; Grinspan et al. 1996; Jin et al. 1993; Michailov et al. 2004; Taveggia et al. 2005; Vartanian et al. 1997). Levels of NRG1 type III correlate with the presence and thickness of myelin as well as the formation of Remak bundles (Michailov et al. 2004; Taveggia et al. 2005). The heterodimers of ErbB2 and ErbB3 mediate signaling through several pathways involving PI3K/Akt, Erk1/2, Ca^{2+}, FAK, and Rac/Cdc42 (Newbern and Birchmeier 2010).

Proteases have been implicated in NRG1–ErbB interactions of axons and Schwann cells. Recent studies have shown that β-amyloid-converting enzyme (BACE1), a β-secretase present in axons, is associated with myelination (Hu et al. 2006; Willem et al. 2006). BACE1-null mice have reduced PNS myelin and remyelination capacity, and consequently exhibit thinner PNS and CNS myelin with reduced levels of myelin proteins. The impaired myelination and remyelination of BACE1-null mice were attributable to reduced rates of NRG1 cleavage, and it is suggested that BACE1 cleaves NRG1 to facilitate its binding to ErbB receptors (Hu et al. 2006). By contrast, downregulation of the axonal α-secretase, tumor necrosis factor-α-converting enzyme (TACE, ADAM17), causes hypermyelination and ectopic myelination that is similar to NRG1 type III overexpression (La Marca et al. 2011). These results indicate that the neuronal α-secretase cleaves NRG1 type III into an inactive form. Another α-secretase, ADAM10, has little effect on myelination, although it can also cleave NRG1 (Freese et al. 2009; Luo et al. 2011). Collectively, behavior of Schwann cells is regulated by NRG1, but this signaling can be modulated by activating and inactivating proteases expressed in neurons.

NRG could also mediate Schwann cell differentiation through axonal neurotrophin signaling in response to neurotrophin release from Schwann cells. It has been suggested that neurotrophins can induce Schwann cell myelination along with increase of axonal diameters (Voyvodic 1989). However, the effect of neurotrophins, such as NGF, is generally restricted to neurons expressing TrkA, and thus it suggests that myelination is facilitated by indirect mechanisms mediated by signals from axons rather than direct glial stimulation by neurotrophin (Rosenberg et al. 2006). The effect of neurotrophins may be mediated by increased NRG1 isoforms, which in turn stimulate the myelination of these DRG axons by Schwann cells (Chan et al. 2004; Esper and Loeb 2004).

Axonal signaling involving proteolytic enzymes may also be required for maintenance of myelin. Expression of prion protein PrPc in axons, but not in Schwann cells, is required for maintenance of the myelin sheath during adulthood (Bremer et al. 2010). Interestingly, a proteolytic cleavage product of PrPc is sufficient to prevent chronic demyelinating polyneuropathy caused by PrPc deficiency (Bremer et al. 2010). the molecular mechanisms of myelin maintenance by axons are still elusive but may have some important implication in the pathophysiology of adult-onset demyelinating diseases.

4.4 Metabolic Link Between Schwann Cells and Axons

Axon-ensheathing cells in the vertebrate nervous system, including Schwann cells, have specific roles for myelination and rapid saltatory conduction, but previous studies have also revealed further roles of these glia in axonal support, in particular, such as survival of the axons that they ensheath (Nave 2010a). This concept is supported by observations that axons are predisposed to degeneration in primary diseases of myelin. For example, recent studies have revealed frequent axonal transections and progressive axon loss in the inflammatory demyelinating disease of CNS, multiple sclerosis (Ferguson et al. 1997; Trapp and Nave 2008; Trapp et al. 1998). Progressive axonal degeneration is also found in human neurological diseases that affect oligodendrocytes, such as leukodystrophies (Nave and Trapp 2008). Inherited peripheral neuropathies, Charcot-Marie-Tooth disease (CMT) type 1, which are caused by Schwann cell dysfunction, also exhibit axon degeneration and loss, which are common among all CMT diseases (Nave et al. 2007). CMT1 is caused by mutations in molecules expressed in Schwann cells, including peripheral myelin protein 22 (*PMP22*) and myelin protein zero (*MPZ*; P_0), which are characterized by demyelination (Nave et al. 2007; Scherer and Wrabetz 2008). Although these diseases exhibit demyelination along with axonal degeneration, it was suggested that myelinating glia support axonal functions independently of myelin. In the CNS, mouse mutants with specific oligodendrocyte defects, such as absence of PLP or CNP, show normal myelin formation but display pathology of progressive axonal loss in the CNS (Griffiths et al. 1998; Lappe-Siefke et al. 2003). In the PNS, some mutations in *MPZ* can cause an axonal form of CMT disease, CMT type 2, where conduction velocity and myelination are not affected but sensory defects and hearing loss are caused by loss of axons (Laura et al. 2007). Typically, the genes causing CMT type 2 are expressed in neurons, but P_0 abundant in PNS myelin is also related to CMT type 2, which is characterized by axonal loss with relatively spared myelin. In mice lacking myelin-associated glycoprotein (MAG), which is expressed by myelinating glia, myelination is normal, but some axons degenerate and axonal diameters are reduced in PNS as well as CNS (Nguyen et al. 2009; Yin et al. 1998).

Axonal degeneration contributes to permanent neurological disability in primary diseases of myelin (Nave et al. 2007; Nave and Trapp 2008; Trapp and Nave 2008). Although the mechanisms of axonal pathology and degeneration after demyelination or dysmyelination are not yet fully understood, they may be associated with the influence of myelin, which alters the structure and metabolism of the axon (de Waegh et al. 1992; Sanchez et al. 1996). One of the influences of myelinating glia is the increase of axonal caliber mediated by posttranslational modifications of axonal cytoskeletons (Colello et al. 1994; Kirkpatrick et al. 2001; Windebank et al. 1985). Signal transduction pathways between axons and myelinating glia would affect posttranslational modification of axonal cytoskeletal proteins including neurofilaments, microtubules, and their associated proteins, which controls axonal caliber and transport (Sousa and Bhat 2007). In *shiverer* mice, where oligodendrocytes form only a few layers of noncompacted myelin around axons without any signs of oligodendrocyte degeneration, the axonal cytoskeletons fail to fully mature and axon diameters remain small (Brady et al. 1999; Griffiths et al. 1998; Inoue et al. 1981; Rosenbluth 1980; Shine et al. 1992). The small axonal diameter in the *shiverer* mutants is caused by narrowly spaced nonphosphorylated neurofilaments and microtubules, which are reminiscent axons in MAG-deficient mice (Nguyen et al. 2009; Yin et al. 1998). However, these axons in *shiverer* mutants do not degenerate. Axons without normal myelin in *shiverer* mice more easily degenerate when oligodendrocytes are further compromised by the absence of PLP1. These results indicate that perturbed maturation of axonal structures itself does not cause axonal degeneration or loss.

The progressive and distally pronounced axonal degeneration in myelin deficit may be related to adaptation and impairment of metabolic homeostasis in axons, given that cell bodies and distal segments of long axons could be distinct biochemical compartments, with respect to metabolic reactions (Nave 2010a).

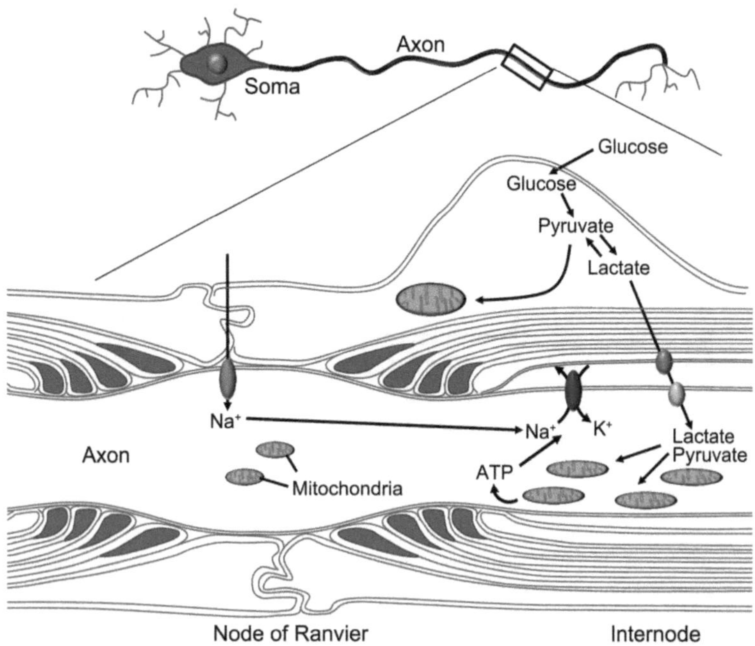

Fig. 4.3 Schematic drawing of metabolic flow upon nerve conduction in myelinated peripheral nerves. Nerve conduction causes Na^+ influx through voltage-gated Na^+ channels at the nodal axolemma. Na^+ is required to be excluded through internodal Na^+/K^+-ATPase in an energy-dependent manner. Axonal mitochondria are enriched in internodal axoplasm, and energy substrates for these mitochondria are likely to be provided through myelinating Schwann cells. Current evidence indicates that lactate generated by Schwann cells is transferred to axons through unidentified transporters

Neuronal Na^+/K^+-ATPases, which use most axonal ATP to exchange axoplasmic Na^+ with extracellular K^+, are present along the entire internodal axolemma (McGrail et al. 1991; Young et al. 2008), suggesting that axonal energy demands on nerve conduction are not restricted around the nodes of Ranvier (Fig. 4.3). This concept is supported by previous observation that the bulk of mitochondrial volume resides in internodes, and the mitochondrial distribution also suggests that most ATP within axons is generated there (Fig. 4.3) (Ohno et al. 2011; Perge et al. 2009). Because mitochondria are the major source of ATP, the internodal enrichment of mitochondria also helps facilitate axonal transport, which is also energy dependent.

Impaired mitochondrial distribution and function have been implicated in the pathogenesis of myelin diseases (Coleman 2005; Trapp and Stys 2009). Axonal conduction/depolarization depends on activation of voltage-gated Na^+ channels. For repetitive conduction, the axolemma must exchange axonal Na^+ for extracellular K^+ in an energy-dependent manner by Na^+/K^+ ATPases. By concentrating voltage-gated Na^+ channels in the nodal axolemma, myelin not only increases the speed of nerve conduction but also conserves energy. Disruption of normal myelin would

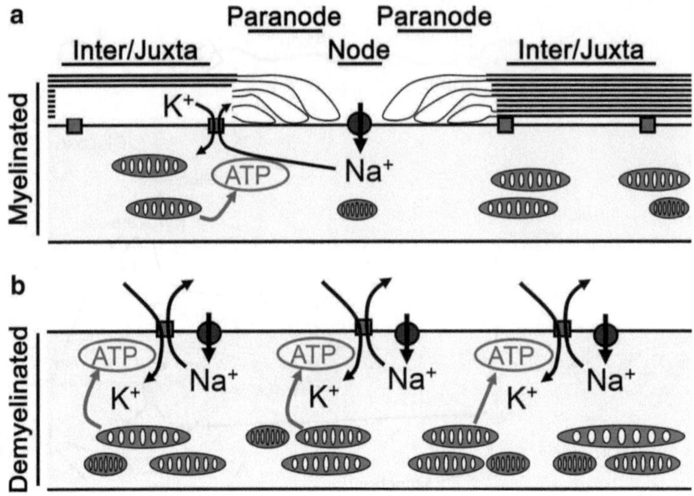

Fig. 4.4 Alterations of energy demand and mitochondrial distribution in myelinated and demyelinated axons. In myelinated axons (**a**), voltage-gated Na^+ channels are concentrated in nodal axolemma. Upon nerve conduction, a limited amount of Na^+ enters the axons, which is excluded from internodal/juxtaparanodal Na^+/K^+-ATPases consuming ATP. These ATP are likely to be originated from axonal mitochondria enriched in intermodal/juxtaparanodal (*Inter/Juxta*) regions. In demyelinated axons (**b**), Na^+ channels redistribute diffusely along the entire axolemma to restore conduction. Upon nerve conduction, more Na^+ gets into the axons, and more ATP is required for Na^+ exclusion. The demyelinated axons have increased volume of axonal mitochondria, presumably to produce ATP sufficient for the Na^+ exclusion

therefore require an adaptive response of energy metabolism from axons. Upon demyelination, mitochondrial volume is increased in the demyelinated axons of human brain and animal models (Mahad et al. 2009; Mutsaers and Carroll 1998; Sathornsumetee et al. 2000; Witte et al. 2009). After demyelination, Na^+ channels diffusely redistribute along the demyelinated axolemma to restore nerve conduction at the expense of increased ATP consumption to drive the Na^+/K^+ ATPases (Craner et al. 2004; Waxman 2008). It is reasonable, therefore, that increases in axonal mitochondrial sizes parallel the increased energy demands of nerve conduction after demyelination in PNS and CNS axons (Fig. 4.4) (Kiryu-Seo et al. 2010; Zambonin et al. 2011). It was also demonstrated that perturbed paranodal junctions in the PNS result in the accumulation of axonal mitochondria around the nodal regions (Einheber et al. 2006; Sun et al. 2009). Dysmyelination increased densities of axonal mitochondria in CNS, as shown in *shiverer* mice and PLP1 mutants (Andrews et al. 2006; Hogan et al. 2009). Collectively, these results support the concept that metabolic relationship between axons and Schwann cells is critically dependent on axo–glial interactions and associated with adaptive alterations in mitochondrial functions.

Molecular mechanisms regulating metabolic adaptation of axonal mitochondria against demyelination or dysmyelination still remain to be elucidated but are likely to be involved in regulation of two populations of axonal mitochondria. The majority of axonal mitochondria are present at stationary sites, which can be composed of single or multiple stationary mitochondria and enriched in axonal areas with high ATP consumption such as growth cones (Kiryu-Seo et al. 2010; Misgeld et al. 2007; Saxton and Hollenbeck 2012; Sheng and Cai 2012). Motile mitochondria are generally smaller and are transported throughout the axon in anterograde and retrograde directions. Demyelination increases the size of the stationary site as well as the transport velocity of axonal mitochondria (Kiryu-Seo et al. 2010). After remyelination, stationary site size and transport velocity were similar to those in the myelinated axons. It was indicated that these mitochondrial alterations were mediated by adaptive responses, at least partly, involving a stress-induced transcription factor, activating transcription factor 3 (ATF3) (Kiryu-Seo et al. 2010). It is also possible that key aspects of this regulation would include increased axoplasmic Ca^{2+} and posttranslational modifications of axoplasmic proteins to halt motile mitochondira, because local inhibition of mitochondrial movement can increase the sizes of stationary mitochondria in axons (Chada and Hollenbeck 2004; Macaskill et al. 2009; Morris and Hollenbeck 1993; Wang and Schwarz 2009). As already described, demyelination increases axoplasmic Na^+ as a result of insufficient ATP production, and increased axoplasmic Na^+ in turn increases axoplasmic Ca^{2+} through reverse operation of Na^+/Ca^{2+} exchangers (Trapp and Stys 2009). Axonal survival would be impaired by the generation of nitric oxide by inflammatory cells, which diffuses into demyelinated axons and contributes to perturbation of mitochondrial ATP generation (Smith and Lassmann 2002; Trapp and Stys 2009). Apart from the acute axonal transection and loss mediated by toxic substances from inflammatory cells, axonal degeneration following demyelination or as a result of dysmyelination is also a chronic process taking months or years to develop (Trapp and Nave 2008). The initial axonal response to demyelination, therefore, reestablishes axonal function and is likely to include changes in mitochondrial distribution, behavior, and life cycles.

Axonal mitochondria have limited lifespans that are dependent on their dynamics and presumably modulated by their overall activity, and thus abnormal mitochondrial dynamics results in impaired axonal integrity (Saxton and Hollenbeck 2012; Sheng and Cai 2012). In neuropathies of PNS, this concept is supported by evidence that molecules regulating mitochondrial fusion/fission and transport, such as mitofusin 2 (Mfn2) and ganglioside-induced differentiation associated protein 1 (GDAP1), are responsible for some forms of CMT (Baxter et al. 2002; Cuesta et al. 2002; Niemann et al. 2005; Zuchner et al. 2004). Newly synthesized mitochondria are largely generated in the neuronal cell body, transported down along the axon, and delivered to stationary sites, where they become fused with stationary mitochondria. Dysfunctional mitochondrial segments are removed from stationary mitochondria through a process called fission, and then transported to the neuronal perikarya, where they are degraded (Saxton and Hollenbeck 2012; Twig et al. 2008). This entire life cycle of axonal mitochondria relying on transcription and translation in cell bodies may render distal axonal segments vulnerable to disruption of energetic homeostasis. The degeneration of axons associated with relevant symptoms such as distally pronounced motor and sensory deficits is progressive and length dependent in primary myelin diseases (Marrosu et al. 1998; Zhou and Griffin 2003); this may explain why a progressive length-dependent loss of axons is commonly observed in peripheral neuropathies and leukodystrophies, first affecting fibers innervating distal regions of extremities in the PNS or the longest spinal tracts in the CNS (Griffin and Watson 1988; Suter and Scherer 2003).

Recent studies suggested that myelin-forming glia provide energy substances such as lactate for the axonal energy production (Nave 2010b). This concept is supported by observations that disruption of monocarboxylic acid transporter 1 (MCT1), which mediates lactate transport from oligodendrocytes to axons for the local energy supply to axons, leads to axonal degeneration in CNS (Funfschilling et al. 2012; Lee et al. 2012). Aberrant axonal degeneration under disruption of MCT1 is likely to be caused by reduced lactate export out of the oligodendroglia. The notion that energy substrates of axonal mitochondria are provided by myelinating glia is consistent with the internodal enrichment of axonal mitochondria, which means that mitochondria are more abundant in axonal regions covered by myelin (Ohno et al. 2011). These observations are also supported by previous findings that trophic support provided by myelin and myelin-forming cells is regulated at the level of individual internodes (Griffiths et al. 1998; Yin et al. 2006). Recent studies suggested that glycogen of Schwann cells in myelinated peripheral nerve fibers provides energy substrates for ensheathed axons during impaired supply of glucose (Brown et al. 2012). It was indicated that lactate is a primary substrate that is generated from glycogen and then shuttled from Schwann cells to axons to maintain axonal ATP stores and excitability. Although the beneficial support of Schwann cell glycogen for unmyelinated axons of Remak bundles was not observed under hypoglycemia (Brown et al. 2012), it is possible that glucose is uptaken largely by Schwann cells in Remak fibers, and the Schwann cells in turn provide lactate as energy substrates to the unmyelinated axons under normal conditions (Vega et al. 1998, 2003). Although it remains to be established if MCT1 is expressed and serves as a shuttling molecule in Schwann cells, these studies provided molecular evidences that energy substrates such as lactate are exported to the extracellular space and taken up by ensheathed axons for energy production by axonal mitochondria (Fig. 4.3).

Glycolysis is assumed to occur throughout the axoplasm. However, if the movement of glycolytic enzymes synthesized in the soma is driven by slow axonal transport, the efficiency of axonal glycolysis could be limited in a length-dependent fashion (Spencer et al. 1979). This possibility suggests that long distal axons may require more metabolic support for mitochondrial energy production than short proximal axons, and this metabolic support could be a trophic function of glia (pyruvate, lactate, or its derivates) for axonal mitochondria in long fiber tracts (Nave 2010b). However, it was recently reported that glyceraldehyde-3-phosphate dehydrogenase (GAPDH) localizes on fast-moving vesicles within axons, and that the glycolytic enzymes located on these vesicles are critical to maintain the high velocities characteristic of fast axonal transport (Zala et al. 2013). It remains to be elucidated if the attachment of glycolytic enzymes to vesicles, mediated by huntingtin, Rab2, and posttranslational modifications of the enzymes themselves, is affected in neurological diseases (Tisdale et al. 2004; Yang et al. 2005; Zala et al. 2013). These studies raised a possibility that modulation of glycolytic energy production in long distal axons is involved in axonal degeneration caused by impaired axonal energy metabolism.

Previous studies support the notion that, apart from lactate, exchange of small metabolites between axons and myelinating glia is extensive and bidirectional. In mice with disrupted mitochondrial metabolism, exclusively seen in Schwann cells by selective depletion of mitochondrial transcription factor A (Tfam-SCKO), pathological features similar to peripheral neuropathies were observed, indicating that peripheral neuropathy occurs secondary to mitochondrial dysfunction of Schwann cells (Viader et al. 2011). Disruption of mitochondria in Schwann cells activates an abnormal integrated stress response, and the actions of heme-regulated inhibitor kinase alter lipid metabolism from fatty acid synthesis toward oxidation (Viader et al. 2013). These changes in the lipid metabolism of Schwann cells deplete myelin lipid components and accumulate acylcarnitines, an intermediate of fatty acid β-oxidation, which is released from Schwann cells and induces axonal degeneration.

Schmidt–Lanterman incisures are a series of funnel-shaped clefts among the compact myelin in the PNS and appear as a series of cytoplasmic openings (Hall and Williams 1970; Peters et al. 1991). Incisures contain connexin 32 (Cx32) to form gap junctions and may have important roles in trafficking of ions and small molecules between inner and outer Schwann cell compartments (Balice-Gordon et al. 1998). This concept is supported by the observation that mutations which impair Cx32 functions cause an X-linked form of CMT, and genetic ablation of Cx32 in mice induces similar pathological phenotypes (Anzini et al. 1997; Bergoffen et al. 1993; Scherer et al. 1998). However, axonal loss or degeneration is mild in these knockout mice (Anzini et al. 1997), indicating that Cx32-dependent gap junctions are redundant for the transport of metabolites to axons.

The same metabolites can play distinct roles in differentiation of myelinating glia in PNS and CNS. In the CNS, blocking of Na^+-dependent action potentials inhibits proliferation of oligodendrocytes and myelination (Barres and Raff 1993; Demerens et al. 1996). It was proposed that axonal electrical activity stimulates ATP release from axons and facilitates oligodendrocyte myelination through cytokines released from astrocytes (Ishibashi et al. 2006). By contrast, in the PNS, the axonal release of ATP perturbs Schwann cell differentiation and myelination through the purinergic P2 receptor (Stevens and Fields 2000).

4.5 Conclusions and Perspectives

Recent advances in genetic techniques, transgenic models, and myelinating cultures during the past decades have already begun to elucidate the molecular and cellular mechanisms by which Schwann cells and axons modulate the behaviors and fates of each other. The list of molecules that are associated with defects of Schwann cells and axons in human PNS diseases is rapidly increasing, and improved biological techniques for tailoring rodent models for in vivo and in vitro manipulation or bioimaging have provided various approaches to further studies of the underlying mechanisms. Future studies are necessary to clarify common features and distinct pathways where ensheathing glia support axonal functions and integrity. Further understanding of the cellular mechanisms for Schwann cell support for axons and axonal signals determining Schwann cell behavior continues to be critical to reveal the physiology of the PNS and also to develop new therapies in peripheral nerve diseases.

References

Andrews H, White K, Thomson C, Edgar J, Bates D, Griffiths I, Turnbull D, Nichols P (2006) Increased axonal mitochondrial activity as an adaptation to myelin deficiency in the Shiverer mouse. J Neurosci Res 83:1533–1539

Anzini P, Neuberg DH, Schachner M, Nelles E, Willecke K, Zielasek J, Toyka KV, Suter U, Martini R (1997) Structural abnormalities and deficient maintenance of peripheral nerve myelin in mice lacking the gap junction protein connexin 32. J Neurosci 17:4545–4551

Balice-Gordon RJ, Bone LJ, Scherer SS (1998) Functional gap junctions in the schwann cell myelin sheath. J Cell Biol 142:1095–1104

Barres BA, Raff MC (1993) Proliferation of oligodendrocyte precursor cells depends on electrical activity in axons. Nature (Lond) 361:258–260

Baxter RV, Ben Othmane K, Rochelle JM, Stajich JE, Hulette C, Dew-Knight S, Hentati F, Ben Hamida M, Bel S, Stenger JE, Gilbert JR, Pericak-Vance MA, Vance JM (2002) Ganglioside-induced differentiation-associated protein-1 is mutant in Charcot-Marie-Tooth disease type 4A/8q21. Nat Genet 30:21–22

Bergoffen J, Scherer SS, Wang S, Scott MO, Bone LJ, Paul DL, Chen K, Lensch MW, Chance PF, Fischbeck KH (1993) Connexin mutations in X-linked Charcot-Marie-Tooth disease. Science 262:2039–2042

Bhat MA, Rios JC, Lu Y, Garcia-Fresco GP, Ching W, St. Martin M, Li J, Einheber S, Chesler M, Rosenbluth J, Salzer JL, Bellen HJ (2001) Axon-glia interactions and the domain organization of myelinated axons requires neurexin IV/Caspr/Paranodin. Neuron 30:369–383

Birchmeier C, Nave KA (2008) Neuregulin-1, a key axonal signal that drives Schwann cell growth and differentiation. Glia 56:1491–1497

Boyle ME, Berglund EO, Murai KK, Weber L, Peles E, Ranscht B (2001) Contactin orchestrates assembly of the septate-like junctions at the paranode in myelinated peripheral nerve. Neuron 30:385–397

Brady ST, Witt AS, Kirkpatrick LL, de Waegh SM, Readhead C, Tu PH, Lee VM (1999) Formation of compact myelin is required for maturation of the axonal cytoskeleton. J Neurosci 19:7278–7288

Bremer J, Baumann F, Tiberi C, Wessig C, Fischer H, Schwarz P, Steele AD, Toyka KV, Nave KA, Weis J, Aguzzi A (2010) Axonal prion protein is required for peripheral myelin maintenance. Nat Neurosci 13:310–318

Brown AM, Evans RD, Black J, Ransom BR (2012) Schwann cell glycogen selectively supports myelinated axon function. Ann Neurol 72:406–418

Bunge RP, Bunge MB, Eldridge CF (1986) Linkage between axonal ensheathment and basal lamina production by Schwann cells. Annu Rev Neurosci 9:305–328

Bunge MB, Bunge RP, Kleitman N, Dean AC (1989) Role of peripheral nerve extracellular matrix in Schwann cell function and in neurite regeneration. Dev Neurosci 11:348–360

Buttermore ED, Dupree JL, Cheng J, An X, Tessarollo L, Bhat MA (2011) The cytoskeletal adaptor protein band 4.1B is required for the maintenance of paranodal axoglial septate junctions in myelinated axons. J Neurosci 31:8013–8024

Carey DJ, Todd MS (1987) Schwann cell myelination in a chemically defined medium: demonstration of a requirement for additives that promote Schwann cell extracellular matrix formation. Brain Res 429:95–102

Carey DJ, Eldridge CF, Cornbrooks CJ, Timpl R, Bunge RP (1983) Biosynthesis of type IV collagen by cultured rat Schwann cells. J Cell Biol 97:473–479

Carroll SL, Miller ML, Frohnert PW, Kim SS, Corbett JA (1997) Expression of neuregulins and their putative receptors, ErbB2 and ErbB3, is induced during Wallerian degeneration. J Neurosci 17:1642–1659

Chada SR, Hollenbeck PJ (2004) Nerve growth factor signaling regulates motility and docking of axonal mitochondria. Curr Biol 14:1272–1276

Chan JR, Watkins TA, Cosgaya JM, Zhang C, Chen L, Reichardt LF, Shooter EM, Barres BA (2004) NGF controls axonal receptivity to myelination by Schwann cells or oligodendrocytes. Neuron 43:183–191

Chan JR, Jolicoeur C, Yamauchi J, Elliott J, Fawcett JP, Ng BK, Cayouette M (2006) The polarity protein Par-3 directly interacts with p75NTR to regulate myelination. Science 314:832–836

Charles P, Tait S, Faivre-Sarrailh C, Barbin G, Gunn-Moore F, Denisenko-Nehrbass N, Guennoc AM, Girault JA, Brophy PJ, Lubetzki C (2002) Neurofascin is a glial receptor for the paranodin/Caspr-contactin axonal complex at the axoglial junction. Curr Biol 12:217–220

Chernousov MA, Yu WM, Chen ZL, Carey DJ, Strickland S (2008) Regulation of Schwann cell function by the extracellular matrix. Glia 56:1498–1507

Cohen JA, Yachnis AT, Arai M, Davis JG, Scherer SS (1992) Expression of the neu proto-oncogene by Schwann cells during peripheral nerve development and Wallerian degeneration. J Neurosci Res 31:622–634

Colello RJ, Pott U, Schwab ME (1994) The role of oligodendrocytes and myelin on axon maturation in the developing rat retinofugal pathway. J Neurosci 14:2594–2605

Coleman M (2005) Axon degeneration mechanisms: commonality amid diversity. Nat Rev Neurosci 6:889–898

Craner MJ, Newcombe J, Black JA, Hartle C, Cuzner ML, Waxman SG (2004) Molecular changes in neurons in multiple sclerosis: altered axonal expression of Nav1.2 and Nav1.6 sodium channels and Na^+/Ca^{2+} exchanger. Proc Natl Acad Sci USA 101:8168–8173

Cuesta A, Pedrola L, Sevilla T, Garcia-Planells J, Chumillas MJ, Mayordomo F, LeGuern E, Marin I, Vilchez JJ, Palau F (2002) The gene encoding ganglioside-induced differentiation-associated protein 1 is mutated in axonal Charcot-Marie-Tooth type 4A disease. Nat Genet 30:22–25

de Waegh SM, Lee VM, Brady ST (1992) Local modulation of neurofilament phosphorylation, axonal caliber, and slow axonal transport by myelinating Schwann cells. Cell 68:451–463

Demerens C, Stankoff B, Logak M, Anglade P, Allinquant B, Couraud F, Zalc B, Lubetzki C (1996) Induction of myelination in the central nervous system by electrical activity. Proc Natl Acad Sci USA 93:9887–9892

Denisenko-Nehrbass N, Oguievetskaia K, Goutebroze L, Galvez T, Yamakawa H, Ohara O, Carnaud M, Girault JA (2003) Protein 4.1B associates with both Caspr/paranodin and Caspr2 at paranodes and juxtaparanodes of myelinated fibres. Eur J Neurosci 17:411–416

Dong Z, Brennan A, Liu N, Yarden Y, Lefkowitz G, Mirsky R, Jessen KR (1995) Neu differentiation factor is a neuron-glia signal and regulates survival, proliferation, and maturation of rat Schwann cell precursors. Neuron 15:585–596

Einheber S, Zanazzi G, Ching W, Scherer S, Milner TA, Peles E, Salzer JL (1997) The axonal membrane protein Caspr, a homologue of neurexin IV, is a component of the septate-like paranodal junctions that assemble during myelination. J Cell Biol 139:1495–1506

Einheber S, Bhat MA, Salzer JL (2006) Disrupted axo-glial junctions result in accumulation of abnormal mitochondria at nodes of ranvier. Neuron Glia Biol 2:165–174

Eldridge CF, Bunge MB, Bunge RP, Wood PM (1987) Differentiation of axon-related Schwann cells in vitro. I. Ascorbic acid regulates basal lamina assembly and myelin formation. J Cell Biol 105:1023–1034

Eshed Y, Feinberg K, Carey DJ, Peles E (2007) Secreted gliomedin is a perinodal matrix component of peripheral nerves. J Cell Biol 177:551–562

Esper RM, Loeb JA (2004) Rapid axoglial signaling mediated by neuregulin and neurotrophic factors. J Neurosci 24:6218–6227

Feinberg K, Eshed-Eisenbach Y, Frechter S, Amor V, Salomon D, Sabanay H, Dupree JL, Grumet M, Brophy PJ, Shrager P, Peles E (2010) A glial signal consisting of gliomedin and NrCAM clusters axonal Na^{+} channels during the formation of nodes of Ranvier. Neuron 65:490–502

Feltri ML, Wrabetz L (2005) Laminins and their receptors in Schwann cells and hereditary neuropathies. J Peripher Nerv Syst 10:128–143

Ferguson B, Matyszak MK, Esiri MM, Perry VH (1997) Axonal damage in acute multiple sclerosis lesions. Brain 120:393–399

Fernandez-Valle C, Fregien N, Wood PM, Bunge MB (1993) Expression of the protein zero myelin gene in axon-related Schwann cells is linked to basal lamina formation. Development (Camb) 119:867–880

Freese C, Garratt AN, Fahrenholz F, Endres K (2009) The effects of alpha-secretase ADAM10 on the proteolysis of neuregulin-1. FEBS J 276:1568–1580

Funfschilling U, Supplie LM, Mahad D, Boretius S, Saab AS, Edgar J, Brinkmann BG, Kassmann CM, Tzvetanova ID, Mobius W, Diaz F, Meijer D, Suter U, Hamprecht B, Sereda MW, Moraes CT, Frahm J, Goebbels S, Nave KA (2012) Glycolytic oligodendrocytes maintain myelin and long-term axonal integrity. Nature (Lond) 485:517–521

Griffin JW, Price DL (1981) Schwann cell and glial responses in beta, beta′-iminodipropionitrile intoxication. I. Schwann cell and oligodendrocyte ingrowths. J Neurocytol 10:995–1007

Griffin JW, Watson DF (1988) Axonal transport in neurological disease. Ann Neurol 23:3–13

Griffiths I, Klugmann M, Anderson T, Yool D, Thomson C, Schwab MH, Schneider A, Zimmermann F, McCulloch M, Nadon N, Nave KA (1998) Axonal swellings and degeneration in mice lacking the major proteolipid of myelin. Science 280:1610–1613

Grinspan JB, Marchionni MA, Reeves M, Coulaloglou M, Scherer SS (1996) Axonal interactions regulate Schwann cell apoptosis in developing peripheral nerve: neuregulin receptors and the role of neuregulins. J Neurosci 16:6107–6118

Hall SM, Williams PL (1970) Studies on the "incisures" of Schmidt and Lanterman. J Cell Sci 6:767–791

Haney CA, Sahenk Z, Li C, Lemmon VP, Roder J, Trapp BD (1999) Heterophilic binding of L1 on unmyelinated sensory axons mediates Schwann cell adhesion and is required for axonal survival. J Cell Biol 146:1173–1184

Hartline DK, Colman DR (2007) Rapid conduction and the evolution of giant axons and myelinated fibers. Curr Biol 17:R29–R35

Hogan V, White K, Edgar J, McGill A, Karim S, McLaughlin M, Griffiths I, Turnbull D, Nichols P (2009) Increase in mitochondrial density within axons and supporting cells in response to demyelination in the Plp1 mouse model. J Neurosci Res 87:452–459

Hoke A, Redett R, Hameed H, Jari R, Zhou C, Li ZB, Griffin JW, Brushart TM (2006) Schwann cells express motor and sensory phenotypes that regulate axon regeneration. J Neurosci 26:9646–9655

Hu X, Hicks CW, He W, Wong P, Macklin WB, Trapp BD, Yan R (2006) Bace1 modulates myelination in the central and peripheral nervous system. Nat Neurosci 9:1520–1525

Inoue Y, Nakamura R, Mikoshiba K, Tsukada Y (1981) Fine structure of the central myelin sheath in the myelin deficient mutant Shiverer mouse, with special reference to the pattern of myelin formation by oligodendroglia. Brain Res 219:85–94

Ishibashi T, Dakin KA, Stevens B, Lee PR, Kozlov SV, Stewart CL, Fields RD (2006) Astrocytes promote myelination in response to electrical impulses. Neuron 49:823–832

Jessen KR, Mirsky R (2005) The origin and development of glial cells in peripheral nerves. Nat Rev Neurosci 6:671–682

Jin JJ, Nikitin A, Rajewsky MF (1993) Schwann cell lineage-specific neu (erbB-2) gene expression in the developing rat nervous system. Cell Growth Differ 4:227–237

Kirkpatrick LL, Witt AS, Payne HR, Shine HD, Brady ST (2001) Changes in microtubule stability and density in myelin-deficient shiverer mouse CNS axons. J Neurosci 21:2288–2297

Kiryu-Seo S, Ohno N, Kidd GJ, Komuro H, Trapp BD (2010) Demyelination increases axonal stationary mitochondrial size and the speed of axonal mitochondrial transport. J Neurosci 30:6658–6666

Klambt C, Hummel T, Granderath S, Schimmelpfeng K (2001) Glial cell development in *Drosophila*. Int J Dev Neurosci 19:373–378

La Marca R, Cerri F, Horiuchi K, Bachi A, Feltri ML, Wrabetz L, Blobel CP, Quattrini A, Salzer JL, Taveggia C (2011) TACE (ADAM17) inhibits Schwann cell myelination. Nat Neurosci 14:857–865

Lappe-Siefke C, Goebbels S, Gravel M, Nicksch E, Lee J, Braun PE, Griffiths IR, Nave KA (2003) Disruption of Cnp1 uncouples oligodendroglial functions in axonal support and myelination. Nat Genet 33:366–374

Laura M, Milani M, Morbin M, Moggio M, Ripolone M, Jann S, Scaioli V, Taroni F, Pareyson D (2007) Rapid progression of late onset axonal Charcot-Marie-Tooth disease associated with a novel MPZ mutation in the extracellular domain. J Neurol Neurosurg Psychiatry 78:1263–1266

Lee Y, Morrison BM, Li Y, Lengacher S, Farah MH, Hoffman PN, Liu Y, Tsingalia A, Jin L, Zhang PW, Pellerin L, Magistretti PJ, Rothstein JD (2012) Oligodendroglia metabolically support axons and contribute to neurodegeneration. Nature (Lond) 487:443–448

Lewallen KA, Shen YA, De la Torre AR, Ng BK, Meijer D, Chan JR (2011) Assessing the role of the cadherin/catenin complex at the Schwann cell–axon interface and in the initiation of myelination. J Neurosci 31:3032–3043

Luo X, Prior M, He W, Hu X, Tang X, Shen W, Yadav S, Kiryu-Seo S, Miller R, Trapp BD, Yan R (2011) Cleavage of neuregulin-1 by BACE1 or ADAM10 protein produces differential effects on myelination. J Biol Chem 286:23967–23974

Macaskill AF, Rinholm JE, Twelvetrees AE, Arancibia-Carcamo IL, Muir J, Fransson A, Aspenstrom P, Attwell D, Kittler JT (2009) Miro1 is a calcium sensor for glutamate receptor-dependent localization of mitochondria at synapses. Neuron 61:541–555

Mackenzie ML, Shorer Z, Ghabriel MN, Allt G (1984) Myelinated nerve fibres and the fate of lanthanum tracer: an in vivo study. J Anat 138:1–14

Mahad DJ, Ziabreva I, Campbell G, Lax N, White K, Hanson PS, Lassmann H, Turnbull DM (2009) Mitochondrial changes within axons in multiple sclerosis. Brain 132:1161–1174

Marrosu MG, Vaccargiu S, Marrosu G, Vannelli A, Cianchetti C, Muntoni F (1998) Charcot-Marie-Tooth disease type 2 associated with mutation of the myelin protein zero gene. Neurology 50:1397–1401

Martini R (1994) Expression and functional roles of neural cell surface molecules and extracellular matrix components during development and regeneration of peripheral nerves. J Neurocytol 23:1–28

Maurel P, Einheber S, Galinska J, Thaker P, Lam I, Rubin MB, Scherer SS, Murakami Y, Gutmann DH, Salzer JL (2007) Nectin-like proteins mediate axon Schwann cell interactions along the internode and are essential for myelination. J Cell Biol 178:861–874

McGrail KM, Phillips JM, Sweadner KJ (1991) Immunofluorescent localization of three Na, K-ATPase isozymes in the rat central nervous system: both neurons and glia can express more than one Na, K-ATPase. J Neurosci 11:381–391

Menegoz M, Gaspar P, Le Bert M, Galvez T, Burgaya F, Palfrey C, Ezan P, Arnos F, Girault JA (1997) Paranodin, a glycoprotein of neuronal paranodal membranes. Neuron 19:319–331

Meyer M, Matsuoka I, Wetmore C, Olson L, Thoenen H (1992) Enhanced synthesis of brain-derived neurotrophic factor in the lesioned peripheral nerve: different mechanisms are responsible for the regulation of BDNF and NGF mRNA. J Cell Biol 119:45–54

Michailov GV, Sereda MW, Brinkmann BG, Fischer TM, Haug B, Birchmeier C, Role L, Lai C, Schwab MH, Nave KA (2004) Axonal neuregulin-1 regulates myelin sheath thickness. Science 304:700–703

Mierzwa A, Shroff S, Rosenbluth J (2011) Permeability of the paranodal junction of myelinated nerve fibers. J Neurosci 30:15962–15968

Mirsky R, Woodhoo A, Parkinson DB, Arthur-Farraj P, Bhaskaran A, Jessen KR (2008) Novel signals controlling embryonic Schwann cell development, myelination and dedifferentiation. J Peripher Nerv Syst 13:122–135

Misgeld T, Kerschensteiner M, Bareyre FM, Burgess RW, Lichtman JW (2007) Imaging axonal transport of mitochondria in vivo. Nat Methods 4:559–561

Morris RL, Hollenbeck PJ (1993) The regulation of bidirectional mitochondrial transport is coordinated with axonal outgrowth. J Cell Sci 104:917–927

Morris JK, Lin W, Hauser C, Marchuk Y, Getman D, Lee KF (1999) Rescue of the cardiac defect in ErbB2 mutant mice reveals essential roles of ErbB2 in peripheral nervous system development. Neuron 23:273–283

Morrissey TK, Levi AD, Nuijens A, Sliwkowski MX, Bunge RP (1995) Axon-induced mitogenesis of human Schwann cells involves heregulin and p185erbB2. Proc Natl Acad Sci USA 92:1431–1435

Mutsaers SE, Carroll WM (1998) Focal accumulation of intra-axonal mitochondria in demyelination of the cat optic nerve. Acta Neuropathol 96:139–143

Nave KA (2010a) Myelination and support of axonal integrity by glia. Nature (Lond) 468:244–252

Nave KA (2010b) Myelination and the trophic support of long axons. Nat Rev Neurosci 11:275–283

Nave KA, Trapp BD (2008) Axon-glial signaling and the glial support of axon function. Annu Rev Neurosci 31:535–561

Nave KA, Sereda MW, Ehrenreich H (2007) Mechanisms of disease: inherited demyelinating neuropathies–from basic to clinical research. Nat Clin Pract Neurol 3:453–464

Newbern J, Birchmeier C (2010) Nrg1/ErbB signaling networks in Schwann cell development and myelination. Semin Cell Dev Biol 21:922–928

Nguyen T, Mehta NR, Conant K, Kim KJ, Jones M, Calabresi PA, Melli G, Hoke A, Schnaar RL, Ming GL, Song H, Keswani SC, Griffin JW (2009) Axonal protective effects of the myelin-associated glycoprotein. J Neurosci 29:630–637

Niemann A, Ruegg M, La Padula V, Schenone A, Suter U (2005) Ganglioside-induced differentiation associated protein 1 is a regulator of the mitochondrial network: new implications for Charcot-Marie-Tooth disease. J Cell Biol 170:1067–1078

Ohno N, Kidd GJ, Mahad D, Kiryu-Seo S, Avishai A, Komuro H, Trapp BD (2011) Myelination and axonal electrical activity modulate the distribution and motility of mitochondria at CNS nodes of Ranvier. J Neurosci 31:7249–7258

Park J, Liu B, Chen T, Li H, Hu X, Gao J, Zhu Y, Zhu Q, Qiang B, Yuan J, Peng X, Qiu M (2008) Disruption of Nectin-like 1 cell adhesion molecule leads to delayed axonal myelination in the CNS. J Neurosci 28:12815–12819

Perge JA, Koch K, Miller R, Sterling P, Balasubramanian V (2009) How the optic nerve allocates space, energy capacity, and information. J Neurosci 29:7917–7928

Perkins GA, Sosinsky GE, Ghassemzadeh S, Perez A, Jones Y, Ellisman MH (2008) Electron tomographic analysis of cytoskeletal cross-bridges in the paranodal region of the node of Ranvier in peripheral nerves. J Struct Biol 161:469–480

Peters A, Palay SL, Webster H (1991) The fine structure of the nervous system: the neurons and supporting cells. Oxford University Press, New York

Podratz JL, Rodriguez E, Windebank AJ (2001) Role of the extracellular matrix in myelination of peripheral nerve. Glia 35:35–40

Poliak S, Gollan L, Martinez R, Custer A, Einheber S, Salzer JL, Trimmer JS, Shrager P, Peles E (1999) Caspr2, a new member of the neurexin superfamily, is localized at the juxtaparanodes of myelinated axons and associates with K^+ channels. Neuron 24:1037–1047

Rasi K, Hurskainen M, Kallio M, Staven S, Sormunen R, Heape AM, Avila RL, Kirschner D, Muona A, Tolonen U, Tanila H, Huhtala P, Soininen R, Pihlajaniemi T (2010) Lack of collagen XV impairs peripheral nerve maturation and, when combined with laminin-411 deficiency, leads to basement membrane abnormalities and sensorimotor dysfunction. J Neurosci 30:14490–14501

Riethmacher D, Sonnenberg-Riethmacher E, Brinkmann V, Yamaai T, Lewin GR, Birchmeier C (1997) Severe neuropathies in mice with targeted mutations in the ErbB3 receptor. Nature (Lond) 389:725–730

Rios JC, Melendez-Vasquez CV, Einheber S, Lustig M, Grumet M, Hemperly J, Peles E, Salzer JL (2000) Contactin-associated protein (Caspr) and contactin form a complex that is targeted to the paranodal junctions during myelination. J Neurosci 20:8354–8364

Robert A, Jirounek P (1994) Uptake of potassium by nonmyelinating Schwann cells induced by axonal activity. J Neurophysiol 72:2570–2579

Rosenberg SS, Ng BK, Chan JR (2006) The quest for remyelination: a new role for neurotrophins and their receptors. Brain Pathol 16:288–294

Rosenbluth J (1980) Central myelin in the mouse mutant *shiverer*. J Comp Neurol 194:639–648

Rosenbluth J (2009) Multiple functions of the paranodal junction of myelinated nerve fibers. J Neurosci Res 87:3250–3258

Salzer JL (2003) Polarized domains of myelinated axons. Neuron 40:297–318

Salzer JL, Brophy PJ, Peles E (2008) Molecular domains of myelinated axons in the peripheral nervous system. Glia 56:1532–1540

Sanchez I, Hassinger L, Paskevich PA, Shine HD, Nixon RA (1996) Oligodendroglia regulate the regional expansion of axon caliber and local accumulation of neurofilaments during development independently of myelin formation. J Neurosci 16:5095–5105

Sathornsumetee S, McGavern DB, Ure DR, Rodriguez M (2000) Quantitative ultrastructural analysis of a single spinal cord demyelinated lesion predicts total lesion load, axonal loss, and neurological dysfunction in a murine model of multiple sclerosis. Am J Pathol 157:1365–1376

Saxton WM, Hollenbeck PJ (2012) The axonal transport of mitochondria. J Cell Sci 125:2095–2104

Scherer SS, Wrabetz L (2008) Molecular mechanisms of inherited demyelinating neuropathies. Glia 56:1578–1589

Scherer SS, Xu YT, Nelles E, Fischbeck K, Willecke K, Bone LJ (1998) Connexin32-null mice develop demyelinating peripheral neuropathy. Glia 24:8–20

Sheng ZH, Cai Q (2012) Mitochondrial transport in neurons: impact on synaptic homeostasis and neurodegeneration. Nat Rev Neurosci 13:77–93

Sherman DL, Tait S, Melrose S, Johnson R, Zonta B, Court FA, Macklin WB, Meek S, Smith AJ, Cottrell DF, Brophy PJ (2005) Neurofascins are required to establish axonal domains for saltatory conduction. Neuron 48:737–742

Shine HD, Readhead C, Popko B, Hood L, Sidman RL (1992) Morphometric analysis of normal, mutant, and transgenic CNS: correlation of myelin basic protein expression to myelinogenesis. J Neurochem 58:342–349

Shroff S, Mierzwa A, Scherer SS, Peles E, Arevalo JC, Chao MV, Rosenbluth J (2011) Paranodal permeability in "myelin mutants". Glia 59:1447–1457

Smith KJ, Lassmann H (2002) The role of nitric oxide in multiple sclerosis. Lancet Neurol 1:232–241

Sousa AD, Bhat MA (2007) Cytoskeletal transition at the paranodes: the Achilles' heel of myelinated axons. Neuron Glia Biol 3:169–178

Spencer PS, Thomas PK (1974) Ultrastructural studies of the dying-back process. II. The sequestration and removal by Schwann cells and oligodendrocytes of organelles from normal and diseases axons. J Neurocytol 3:763–783

Spencer PS, Sabri MI, Schaumburg HH, Moore CL (1979) Does a defect of energy metabolism in the nerve fiber underlie axonal degeneration in polyneuropathies? Ann Neurol 5:501–507

Spiegel I, Adamsky K, Eshed Y, Milo R, Sabanay H, Sarig-Nadir O, Horresh I, Scherer SS, Rasband MN, Peles E (2007) A central role for Necl4 (SynCAM4) in Schwann cell–axon interaction and myelination. Nat Neurosci 10:861–869

Stevens B, Fields RD (2000) Response of Schwann cells to action potentials in development. Science 287:2267–2271

Sun XY, Takagishi Y, Okabe E, Chishima Y, Kanou Y, Murase S, Mizumura K, Inaba M, Komatsu Y, Hayashi Y, Peles E, Oda S, Murata Y (2009) A novel Caspr mutation causes the shambling

mouse phenotype by disrupting axoglial interactions of myelinated nerves. J Neuropathol Exp Neurol 68:1207–1218
Suter U, Scherer SS (2003) Disease mechanisms in inherited neuropathies. Nat Rev Neurosci 4:714–726
Tait S, Gunn-Moore F, Collinson JM, Huang J, Lubetzki C, Pedraza L, Sherman DL, Colman DR, Brophy PJ (2000) An oligodendrocyte cell adhesion molecule at the site of assembly of the paranodal axo-glial junction. J Cell Biol 150:657–666
Taveggia C, Zanazzi G, Petrylak A, Yano H, Rosenbluth J, Einheber S, Xu X, Esper RM, Loeb JA, Shrager P, Chao MV, Falls DL, Role L, Salzer JL (2005) Neuregulin-1 type III determines the ensheathment fate of axons. Neuron 47:681–694
Tisdale EJ, Kelly C, Artalejo CR (2004) Glyceraldehyde-3-phosphate dehydrogenase interacts with Rab2 and plays an essential role in endoplasmic reticulum to Golgi transport exclusive of its glycolytic activity. J Biol Chem 279:54046–54052
Trachtenberg JT, Thompson WJ (1996) Schwann cell apoptosis at developing neuromuscular junctions is regulated by glial growth factor. Nature (Lond) 379:174–177
Traka M, Goutebroze L, Denisenko N, Bessa M, Nifli A, Havaki S, Iwakura Y, Fukamauchi F, Watanabe K, Soliven B, Girault JA, Karagogeos D (2003) Association of TAG-1 with Caspr2 is essential for the molecular organization of juxtaparanodal regions of myelinated fibers. J Cell Biol 162:1161–1172
Trapp BD, Nave KA (2008) Multiple sclerosis: an immune or neurodegenerative disorder? Annu Rev Neurosci 31:247–269
Trapp BD, Stys PK (2009) Virtual hypoxia and chronic necrosis of demyelinated axons in multiple sclerosis. Lancet Neurol 8:280–291
Trapp BD, Peterson J, Ransohoff RM, Rudick R, Mork S, Bo L (1998) Axonal transection in the lesions of multiple sclerosis. N Engl J Med 338:278–285
Twig G, Elorza A, Molina AJ, Mohamed H, Wikstrom JD, Walzer G, Stiles L, Haigh SE, Katz S, Las G, Alroy J, Wu M, Py BF, Yuan J, Deeney JT, Corkey BE, Shirihai OS (2008) Fission and selective fusion govern mitochondrial segregation and elimination by autophagy. EMBO J 27:433–446
Vartanian T, Goodearl A, Viehover A, Fischbach G (1997) Axonal neuregulin signals cells of the oligodendrocyte lineage through activation of HER4 and Schwann cells through HER2 and HER3. J Cell Biol 137:211–220
Vega C, Poitry-Yamate CL, Jirounek P, Tsacopoulos M, Coles JA (1998) Lactate is released and taken up by isolated rabbit vagus nerve during aerobic metabolism. J Neurochem 71:330–337
Vega C, Martiel JL, Drouhault D, Burckhart MF, Coles JA (2003) Uptake of locally applied deoxyglucose, glucose and lactate by axons and Schwann cells of rat vagus nerve. J Physiol 546:551–564
Viader A, Golden JP, Baloh RH, Schmidt RE, Hunter DA, Milbrandt J (2011) Schwann cell mitochondrial metabolism supports long-term axonal survival and peripheral nerve function. J Neurosci 31:10128–10140
Viader A, Sasaki Y, Kim S, Strickland A, Workman CS, Yang K, Gross RW, Milbrandt J (2013) Aberrant Schwann cell lipid metabolism linked to mitochondrial deficits leads to axon degeneration and neuropathy. Neuron 77:886–898
Voyvodic JT (1989) Target size regulates calibre and myelination of sympathetic axons. Nature (Lond) 342:430–433
Wang X, Schwarz TL (2009) The mechanism of Ca^{2+}-dependent regulation of kinesin-mediated mitochondrial motility. Cell 136:163–174
Waxman SG (2008) Mechanisms of disease: sodium channels and neuroprotection in multiple sclerosis—current status. Nat Clin Pract Neurol 4:159–169
Webster Hde F (1971) Myelinogenesis. Structural aspects. Neurosci Res Program Bull 9:470–477
Willem M, Garratt AN, Novak B, Citron M, Kaufmann S, Rittger A, DeStrooper B, Saftig P, Birchmeier C, Haass C (2006) Control of peripheral nerve myelination by the beta-secretase BACE1. Science 314:664–666

Windebank AJ, Wood P, Bunge RP, Dyck PJ (1985) Myelination determines the caliber of dorsal root ganglion neurons in culture. J Neurosci 5:1563–1569

Witte ME, Bo L, Rodenburg RJ, Belien JA, Musters R, Hazes T, Wintjes LT, Smeitink JA, Geurts JJ, De Vries HE, van der Valk P, van Horssen J (2009) Enhanced number and activity of mitochondria in multiple sclerosis lesions. J Pathol 219:193–204

Woldeyesus MT, Britsch S, Riethmacher D, Xu L, Sonnenberg-Riethmacher E, Abou-Rebyeh F, Harvey R, Caroni P, Birchmeier C (1999) Peripheral nervous system defects in erbB2 mutants following genetic rescue of heart development. Genes Dev 13:2538–2548

Xiao J, Kilpatrick TJ, Murray SS (2009) The role of neurotrophins in the regulation of myelin development. Neurosignals 17:265–276

Yang J, Gibson B, Snider J, Jenkins CM, Han X, Gross RW (2005) Submicromolar concentrations of palmitoyl-CoA specifically thioesterify cysteine 244 in glyceraldehyde-3-phosphate dehydrogenase inhibiting enzyme activity: a novel mechanism potentially underlying fatty acid induced insulin resistance. Biochemistry 44:11903–11912

Yin X, Crawford TO, Griffin JW, Tu P, Lee VM, Li C, Roder J, Trapp BD (1998) Myelin-associated glycoprotein is a myelin signal that modulates the caliber of myelinated axons. J Neurosci 18:1953–1962

Yin X, Baek RC, Kirschner DA, Peterson A, Fujii Y, Nave KA, Macklin WB, Trapp BD (2006) Evolution of a neuroprotective function of central nervous system myelin. J Cell Biol 172:469–478

Young EA, Fowler CD, Kidd GJ, Chang A, Rudick R, Fisher E, Trapp BD (2008) Imaging correlates of decreased axonal Na^+/K^+ ATPase in chronic multiple sclerosis lesions. Ann Neurol 63:428–435

Zala D, Hinckelmann MV, Yu H, Lyra da Cunha MM, Liot G, Cordelieres FP, Marco S, Saudou F (2013) Vesicular glycolysis provides on-board energy for fast axonal transport. Cell 152:479–491

Zambonin JL, Zhao C, Ohno N, Campbell GR, Engeham S, Ziabreva I, Schwarz N, Lee SE, Frischer JM, Turnbull DM, Trapp BD, Lassmann H, Franklin RJ, Mahad DJ (2011) Increased mitochondrial content in remyelinated axons: implications for multiple sclerosis. Brain 134:1901–1913

Zhou L, Griffin JW (2003) Demyelinating neuropathies. Curr Opin Neurol 16:307–313

Zuchner S, Mersiyanova IV, Muglia M, Bissar-Tadmouri N, Rochelle J, Dadali EL, Zappia M, Nelis E, Patitucci A, Senderek J, Parman Y, Evgrafov O, Jonghe PD, Takahashi Y, Tsuji S, Pericak-Vance MA, Quattrone A, Battaloglu E, Polyakov AV, Timmerman V, Schroder JM, Vance JM (2004) Mutations in the mitochondrial GTPase mitofusin 2 cause Charcot-Marie-Tooth neuropathy type 2A. Nat Genet 36:449–451

Chapter 5
Schwann Cell-Dependent Regulation of Peripheral Nerve Injury and Repair

Keiichiro Susuki

Abstract Schwann cells play significant roles in multiple aspects of peripheral nerve structures and functions in both health and disease. Recent studies have uncovered how Schwann cells respond to peripheral nerve injury. Schwann cells trans-differentiate to distinct phenotypes after nerve injury and contribute to myelin debris clearance, attract macrophages into the lesions, protect injured neurons, promote axonal regrowth, and finally remyelinate and reform nodes of Ranvier along the regenerated axons. Understanding the molecular mechanisms behind the Schwann cell repair phenotype will provide important clues to establish specific treatment strategies to protect peripheral nerves from injury and facilitate functional recovery.

Keywords Axonal degeneration • Injury • Myelin • Node of Ranvier • Remyelination • Repair • Schwann cell

Abbreviations

Nav	Voltage-gated Na^+ channels
Necl	Nectin-like protein
NRG1	Neuregulin-1
PNS	Peripheral nervous system

K. Susuki (✉)
Department of Neuroscience, Baylor College of Medicine,
One Baylor Plaza, Room S611-MS: BCM295, Houston, TX 77030, USA

Department of Neuroscience, Cell Biology, and Physiology,
Wright State University 3640 Colonel Glenn Hwy., Dayton, OH 45435, USA
e-mail: susuki@bcm.edu

K. Sango and J. Yamauchi (eds.), *Schwann Cell Development and Pathology*,
DOI 10.1007/978-4-431-54764-8_5, © Springer Japan 2014

5.1 Introduction

The main function of the peripheral nervous system (PNS) is to transmit information from the brain to the body and from the body to the brain. Schwann cells are required for rapid and efficient propagation of the action potentials along the long distances of PNS that can exceed 1 m in length. One Schwann cell forms a single myelin sheath, a concentrically laminated structure wrapping around an axon (Jessen and Mirsky 2005; Sherman and Brophy 2005; Pereira et al. 2012). The myelin lamella is formed by the fusion of the apposed inner leaflets of the plasma membrane in Schwann cells (with no intervening cytoplasm). Myelin is rich in lipids (approximately 80 %) and can therefore act as an insulator with high transverse resistance and a low electrical capacitance. The length of the myelin sheath is approximately 1 mm in the PNS, and myelinated axons are ensheathed along their entire length. Between two adjacent myelin segments, there are approximately 1-μm-long gaps called the nodes of Ranvier. Another important role of myelinating Schwann cells is to promote clustering of voltage-gated Na^+ (Nav) channels at the nodal axolemma (see Chap. 8) (Poliak and Peles 2003; Susuki and Rasband 2008). Nav channels at the nodes are responsible for the regeneration of the action potentials, whereas the axons between nodes, or internodal segments, are insulated by myelin sheaths, allowing the rapid and efficient saltatory conduction of the action potentials. Thus, myelination can greatly increase the speed of nerve conduction. Indeed, the evolution of myelin allowed vertebrates to achieve efficient information networks despite their large body size.

The importance of myelinating Schwann cells is further underscored by a wide variety of neurological diseases involving myelinated nerve fibers. For example, in demyelinating forms of Charcot-Marie-Tooth disease (hereditary; see Chap. 6) or Guillain-Barré syndrome (immune mediated; see Chap. 8), the primary lesion is peripheral nerve myelin. Damage to Schwann cells causes loss of axon insulation, and inhibition of action potential regeneration by disruption of nodal Nav channel clusters, thereby disturbing nerve conduction. In addition, because the Schwann cells also contribute to support axonal integrity, severe demyelination results in axonal degeneration. Furthermore, damage to the axons such as traumatic PNS injury causes loss of myelin from disruption of the axon–Schwann cell interaction. However, the peripheral nerves have great capacity to regenerate after injury, because of a permissive environment for regeneration, the intrinsic growth capacity of neurons, and the reprogramming ability of Schwann cells (Chen et al. 2007; Raivich and Makwana 2007; Jessen and Mirsky 2008). In addition to the characteristics in myelination, the Schwann cells play crucial roles during nerve injury and repair process. After peripheral nerve injury, myelin-forming Schwann cells switch to the distinct phenotype to protect injured neurons, clear myelin debris, promote axon regeneration, and finally remyelinate along the regenerated axons (Arthur-Farraj et al. 2012). Despite the similarity in many ways, the remyelination process does not simply recapitulate myelination during early development. This chapter focuses on recent experimental approaches to dissect molecular mechanisms of Schwann cells regulating peripheral nerve degeneration and regeneration.

5.2 Traumatic PNS Injury and Repair

Peripheral nerve injury triggers axonal degeneration in the distal nerve stump. This complex process includes axon and myelin breakdown, changes in the permeability of the blood vessels, Schwann cell proliferation, cytokine network production, and the phagocytosis of myelin fragments by Schwann cells and macrophages (for details, see Chen et al. 2007; Raivich and Makwana 2007; Rotshenker 2011). The experimental model of sciatic nerve crush injury is one of the most popular methods used for the study of peripheral nerve degeneration and regeneration. Crushing sciatic nerves by fine forceps produces abrupt tissue damage at the lesion site accompanied by nerve conduction block (Fig. 5.1a). In the distal region of the sciatic nerves, most axons undergo degeneration by 1 week after crush, although they did not encounter the direct physical impact (Fig. 5.1b, c). The degeneration stops at the first internode in the axons proximal to the lesion site if the injury is mild. By loss of contact with axons, Schwann cells break down myelin and proliferate to facilitate nerve repair. The functional axonal domains including nodes of Ranvier are also almost completely lost by 1 week after crush (Nakata et al. 2008). Macrophages are recruited to the injury sites and contribute to myelin debris clearance together with Schwann cells. Removal of degenerated myelin is critical for repair because PNS myelin contains molecules such as myelin-associated glycoprotein that inhibit axon regeneration (Schäfer et al. 1996; Shen et al. 1998).

Injured axons regenerate readily after crush. Axon injury activates intrinsic growth capacity within the affected neurons to promote axonal regeneration (for details, see Chen et al. 2007). Furthermore, the distal segment of the injured sciatic nerve provides a supportive environment for regeneration. The injured axons start to regrow within a period of weeks and successfully reinnervate the appropriate targets. Functional recovery also requires successful remyelination and node of Ranvier formation along the regenerated axons. To complete the repair process, the Schwann cells contact the regrowing axons and start remyelination (Fig. 5.1c). Reformed myelin segments are characterized by thinner myelin sheaths and shorter internodes, resulting in slower nerve conduction velocity (Nakata et al. 2008). The myelin sheaths become thicker over time, but they are still thinner than uninjured myelin even at 20 weeks after crush. Nodes of Ranvier are reformed concurrently with remyelination. Two weeks after crush, the density of Nav channel clusters is dramatically increased to approximately threefold as compared to uninjured nerves (Nakata et al. 2008). The number tends to decrease over time, although it is still significantly higher than control even at 20 weeks after crush, presumably because of the shorter internodes in regenerated nerves. The process of nodal reformation along the regenerated axons is similar to that during early development (Eshed et al. 2005; Feinberg et al. 2010). Nav channels are first trapped at the edge of Schwann cells by extracellular matrix molecules. The junctions form between axons and Schwann cells at paranodes, restricting the mobility of Nav channels and providing a second mechanism for node assembly. Then, the two Nav channel clusters accumulated at the edges of approaching Schwann cells fuse with each other to form a mature node of Ranvier (Fig. 5.1d).

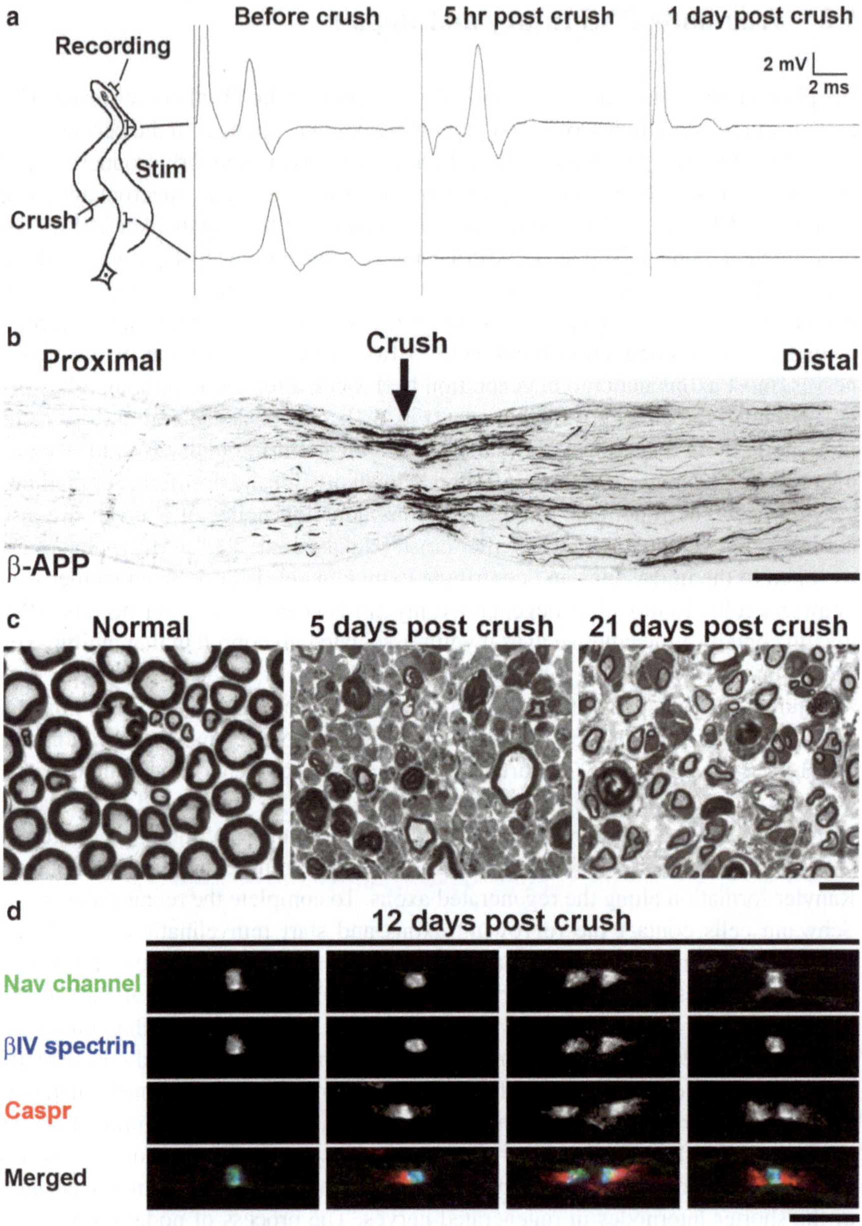

Fig. 5.1 Crush injury in rodent sciatic nerves. **a** Serial nerve conduction study in sciatic nerve from one mouse. The sciatic nerve was crushed halfway between the knee and sciatic notch. The nerve is stimulated at ankle or sciatic notch, and the compound muscle action potentials are recorded from plantar muscle (*top*, ankle stimulation; *bottom*, sciatic notch stimulation). Before crush, no apparent difference is seen between the waveforms after stimulation at the ankle and sciatic notch. Five hours after crush, the action potential is not elicited after stimulation at the sciatic notch, whereas it is preserved after ankle stimulation, suggesting the presence of complete nerve conduction block at the crush site. One day after crush, the amplitude after ankle stimulation

5.3 Repair Phenotype of Schwann Cells

The myelination status of Schwann cells is determined by the balance between positive and negative transcriptional regulators (reviewed by Jessen and Mirsky 2008). During early myelination, positive regulators such as a transcription factor Krox-20 are upregulated in myelinating Schwann cells whereas negative regulators are suppressed. When peripheral nerves are injured, the balance shifts to negative regulators, and the Schwann cells represent the distinct repair phenotype (Fig. 5.2). Transition of Schwann cells to the repair phenotype, subsequent proliferation, and redifferentiation are regulated by multiple players including transcription factors, extracellular matrix proteins, neurotrophic factors, and hormones (Chen et al. 2007; Jessen and Mirsky 2008). Among these molecules, a key to activate the Schwann cell repair program in response to injury is the transcription factor c-Jun (Arthur-Farraj et al. 2012). When Schwann cells lose contact with axons in injured nerves, they lose their differentiated morphology. In these cells, c-Jun is rapidly upregulated and a large number of genes related to myelination are downregulated. Furthermore, these Schwann cells upregulate a group of molecules that are normally found in immature cells before myelination and reenter the cell cycle. Because of similarities to the immature Schwann cells, this process has been termed dedifferentiation. However, there are some differences in molecular expression between immature Schwann cells during early myelination and adult denervated Schwann cells (Jessen and Mirsky 2008). The breakdown of myelin sheaths is carried out by the Schwann cells as well as the invading macrophages. Schwann cells that form close contacts with axons are the first among nonneuronal cells to respond to axon injury by rapidly producing cytokines and chemokines and recruiting blood-borne macrophages (Rotshenker 2011). In addition to attracting macrophages to the injured sites, the Schwann cells themselves phagocytose myelin breakdown products and damaged axons. These events are important for preparing an environment that promotes regeneration after injury.

Schwann cells exhibiting the repair phenotype have a significant impact on injured neurons. Most importantly, Schwann cells maintain the neuronal population to ensure effective recovery of the nerves. In mice with selective inactivation of c-Jun in Schwann cells (produced by crossing floxed *c-jun* mice with a mouse Cre

---

Fig. 5.1 (continued) is remarkably reduced, suggesting that the axonal degeneration is extending to the ankle level. **b** Longitudinal sections of mouse tibial branch of sciatic nerve 1 day post crush. Antibodies to β-APP show degenerated axons around crush site and extending toward the distal part of the nerve. *Bar* 0.5 mm. **c** Semithin cross sections of rat sciatic nerves stained with *toluidine blue*. Compared to the uninjured nerve (*left panel*), most of the nerve fibers are degenerated at 5 days post crush (*middle panel*). During regeneration at 21 days post crush (*right panel*), some axons are thinly myelinated. *Bar* 10 μm. **d** Longitudinal sections of mouse sciatic nerve at 12 days post crush. Sections are immunostained by antibodies to nodal proteins Nav channel (*green*), βIV spectrin (*blue*), and paranodal protein Caspr (contactin-associated protein, *red*). Different stages of node formation are depicted: node not associated with paranode, heminode (paranode only one side), binary nodal clusters, and mature node (from *left* to *right*). *Bar* 10 μm

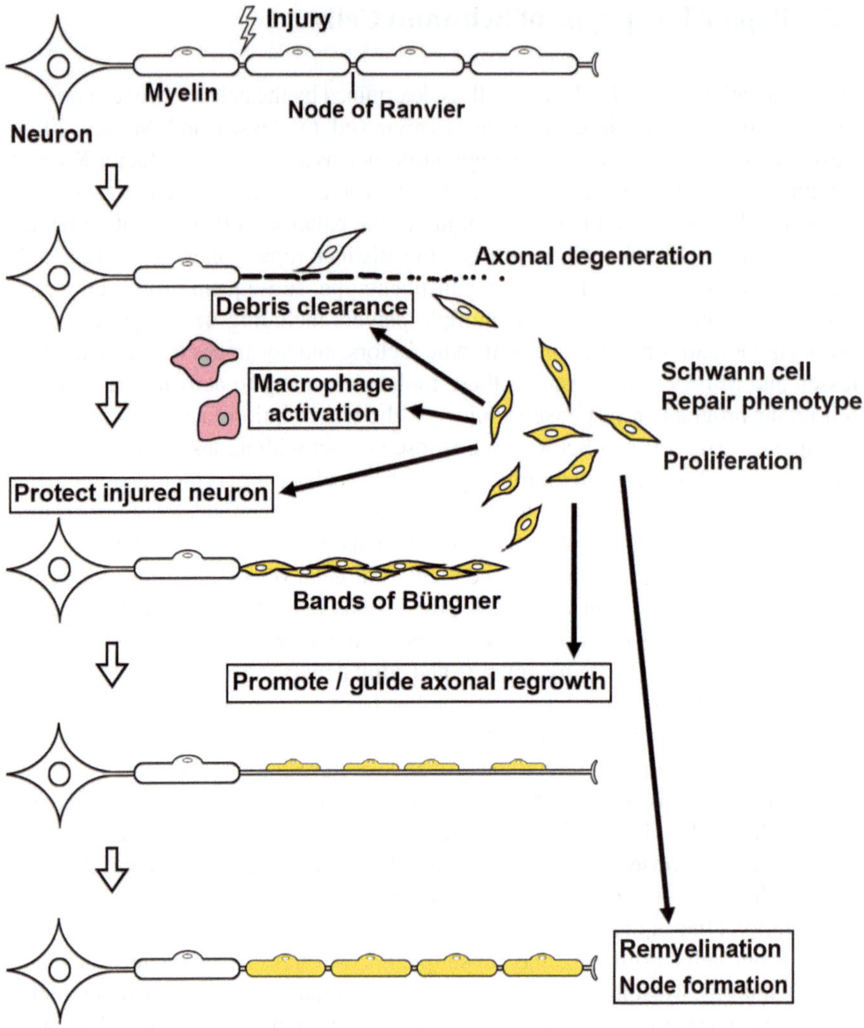

Fig. 5.2 Role of Schwann cells during peripheral nerve degeneration and regeneration. Cartoon shows degeneration and regeneration of myelinated nerve fibers after peripheral nervous system (PNS) traumatic injury. Upon nerve injury, the distal parts of the axons undergo degeneration. By loss of contact with axons, the Schwann cells trans-differentiate to the repair phenotype. These Schwann cells proliferate and contribute to debris clearance, macrophage activation, protecting injured neurons, promoting and guidance of axonal regrowth, and finally remyelination and node of Ranvier formation

line under the control of the Schwann cell-specific promotor myelin protein zero), sciatic nerve crush injury resulted in extensive death of sensory neurons, although ventral horn motoneurons were unchanged (Arthur-Farraj et al. 2012). In another study using the same mutant animals, the absence of c-Jun specifically in Schwann cells severely increased motoneuron death after transection of the facial nerve

(Fontana et al. 2012). These findings demonstrate the roles of Schwann cells in promoting neuronal survival after injury. Furthermore, the Schwann cells also promote axon regrowth. The denervated Schwann cells proliferate and line the endoneurial tubes formed by the remaining basal lamina (bands of Büngner). These uninterrupted tracts of Schwann cells provide a permissive environment for axon regeneration and guide axons back to their targets (Chen et al. 2007; Arthur-Farraj et al. 2012). Following the loss of c-Jun in Schwann cells, the bands of Büngner are not formed after sciatic nerve crush injury, resulting in severely reduced axonal regeneration (Arthur-Farraj et al. 2012). Moreover, c-Jun controls the ability of Schwann cells to support motoneuron survival and axonal regeneration by upregulating genes encoding glial-derived neurotrophic factor and artemin, both of which are ligands for the Ret receptor tyrosine kinase in neurons and have well-described functions in neuronal survival and axonal elongation (Fontana et al. 2012). Taken together, these findings support the notion that Schwann cells have crucial roles to control the injury and repair processes in the PNS. Schwann cell injury response has much in common with trans-differentiation rather than dedifferentiation, because it represents the generation of a distinct Schwann cell repair phenotype, specialized for supporting axonal regrowth and neuronal survival (Arthur-Farraj et al. 2012).

5.4 Remyelination

The final step of functional recovery is remyelination along the regenerated axons. To complete the repair process, Schwann cells envelop the regenerated axons and redifferentiate to form myelin sheaths (Chen et al. 2007). When the Schwann cells recontact the regenerating axons, the expression of a transcription factor Oct-6 is transiently upregulated, indicating that they redifferentiate to a similar state as premyelinating nascent Schwann cells. Subsequently, a transcription factor Krox-20, the main positive regulator for myelination, is expressed and myelin-specific genes are induced as in normal development.

Recent studies utilize the combination of modern molecular biology techniques and sciatic nerve injury models to shed light on the molecules that play key roles in remyelination, although only a limited number of molecules involving myelination have been evaluated in the context of repair. For example, during early development, axonally derived neuregulin 1 (NRG1) type III regulates Schwann cell proliferation, differentiation, and myelination by inducing heterodimerization of the receptors ErbB2 and ErbB3 in Schwann cells and triggering multiple downstream signaling pathways (Nave and Salzer 2006; Taveggia et al. 2010). Similarly, signaling pathways regulated by axonal NRG1 are important for remyelination after nerve injury. In mutant mice with conditionally induced ablation of NRG1 in a subset of adult neurons, axons lacking NRG1 are able to maintain a myelin sheath but are severely impaired in remyelination after sciatic nerve crush injury (Fricker et al. 2011). In transgenic mice overexpressing NRG1 type III in neurons, remyelination after sciatic nerve crush injury was more efficient than wild type (Stassart et al. 2013).

The loss of Erbin, a binding protein that specifically interacts with the ErbB2, attenuates the elevation of ErbB2/3 and NRG1 type III and impaired remyelination after sciatic nerve crush injury (Liang et al. 2012). In addition, NRG1 type I is expressed in Schwann cells during the early time points after crush injury when axons are lost by degeneration and thus axonal NRG1 type III is absent. Specific ablation of NRG1 in Schwann cells did not affect primary myelination and myelin maintenance but strongly impaired remyelination after crush injury. These findings support the idea that the loss of axonal contact triggers denervated Schwann cells to transiently express NRG1 type I as an autocrine/paracrine signal to promote their differentiation and remyelination (Stassart et al. 2013).

In addition to NRG1 and ErbB2/3 signaling, axon–Schwann cell interactions are also mediated by interaction of cell adhesion molecules called nectin-like proteins (Necls) to initiate myelination (Spiegel et al. 2007; Maurel et al. 2007). During early development, Necl4 is expressed in the myelinating Schwann cells at the contact site with axons, and interacts with Necl1 expressed on the axolemma. The role of Necl1–Necl4 interaction in remyelination was demonstrated using a focal demyelination model induced by intraneural lysolecithin injection into rat sciatic nerves (Spiegel et al. 2007). At a time when Schwann cells actively remyelinate, injected soluble Necl4-Fc protein competed with endogenous Schwann cell Necl4 and inhibited remyelination. Polarized localization of Schwann cell Necl4 at the axon-glial interface is stabilized by Schwann cell submembranous cytoskeletal proteins αII and βII spectrins by linking Necl4 to the actin cytoskeleton (Susuki et al. 2011). Silencing expression of spectrins by shRNA in myelinating Schwann cells inhibited myelination in rat dorsal root ganglion neurons and Schwann cell co-culture, and remyelination after crush injury in rat sciatic nerves (Fig. 5.3). Thus, spectrins in Schwann cells integrate the neuron–glia interactions mediated by Necls into the actin-dependent cytoskeletal rearrangements necessary for myelin formation along the regenerated axons.

5.5 Conclusion

This chapter has reviewed the recent progress in research to uncover the molecular mechanisms of traumatic peripheral nerve injury, focusing on the roles of Schwann cells. However, the degeneration and regeneration processes are still poorly understood, and no treatments are available to directly protect myelinated nerve fibers or facilitate repair. How can we attenuate the disease or injury process on peripheral nerves? What is the treatment strategy to achieve functional recovery? As reviewed here, a key to answer these questions is the Schwann cell repair phenotype. During peripheral nerve traumatic injury and subsequent repair process, Schwann cells play significant roles in the clearance of myelin debris, protecting injured neurons, promoting axon regrowth and remyelination along the regenerated axons. Recent molecular biology techniques in combination with nerve injury models allow us to explore the molecular mechanisms in various cell types including Schwann cells

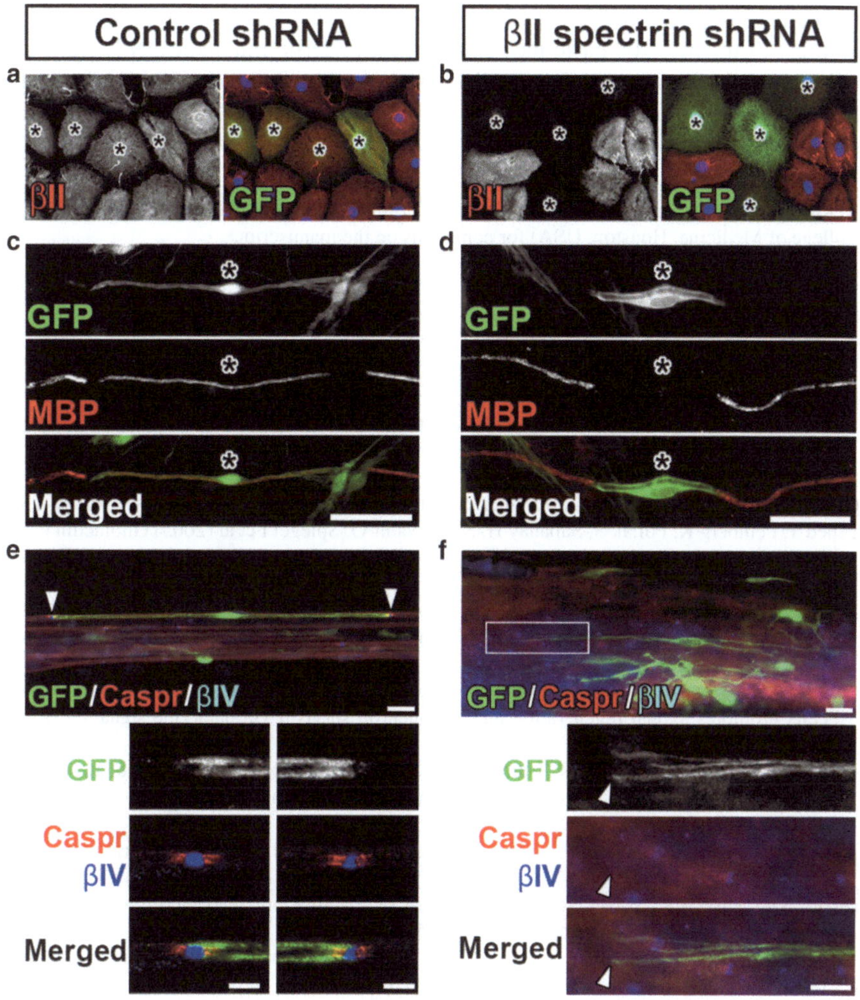

Fig. 5.3 Inhibition of myelination by silencing expression of βII spectrin in Schwann cells. **a, b** Cultured rat Schwann cells 7 days after infection with adenovirus containing control (**a**) or βII spectrin (**b**) shRNAs. Infected cells are identified by GFP (*asterisks*). *Bars* 50 μm. **c, d** Myelinating co-culture of rat dorsal root ganglion neurons and Schwann cells with control (**c**) or βII spectrin (**d**) shRNAs. To visualize myelin sheaths, cells are immunostained for myelin basic protein (MBP) in *red*. *Asterisks* indicate infected cells with GFP expression. *Bars* 50 μm. **e, f** βII spectrin silencing during sciatic nerve regeneration. One day post crush, the adenoviruses with shRNAs were injected into sciatic nerves distal to the injury site. **e** A myelin segment produced by a Schwann cell with control shRNA at 21 days post crush. The nodes (βIV spectrin in *blue*) and paranodes (Caspr in *red*) on both sides (*arrowheads*) are enlarged in *lower panels*. **f** Considerable GFP expression in cells not producing myelin in sciatic nerves injected with βII spectrin shRNA at 21 days post crush. The *boxed area* in *upper panel* is enlarged in the *lower panels*. The *arrowhead* indicates the tip of GFP-positive cells with no overlapping nodes or paranodes. *Bars* 20 μm (**e, f,** *upper panels*); 5 μm (*lower panels* in **e**); 10 μm (*lower panels* in **f**). (Reproduced from Susuki et al. 2011 with modifications)

regulating peripheral nerve degeneration and regeneration. Future research of this nature will uncover the basic mechanisms that underlie traumatic PNS injury and will provide important clues to establish novel therapeutic approaches. Furthermore, those research accomplishments will be applicable for a wide variety of currently intractable PNS diseases.

Acknowledgments The author thanks Daniel R. Zollinger and Dr. Matthew N. Rasband (Baylor College of Medicine, Houston, USA) for comments on the manuscript.

References

Arthur-Farraj PJ, Latouche M, Wilton DK, Quintes S, Chabrol E, Banerjee A et al (2012) c-Jun reprograms Schwann cells of injured nerves to generate a repair cell essential for regeneration. Neuron 75:633–647

Chen ZL, Yu WM, Strickland S (2007) Peripheral regeneration. Annu Rev Neurosci 30:209–233

Eshed Y, Feinberg K, Poliak S, Sabanay H, Sarig-Nadir O, Spiegel I et al (2005) Gliomedin mediates Schwann cell–axon interaction and the molecular assembly of the nodes of Ranvier. Neuron 47:215–229

Feinberg K, Eshed-Eisenbach Y, Frechter S, Amor V, Salomon D, Sabanay H et al (2010) A glial signal consisting of gliomedin and NrCAM clusters axonal Na^+ channels during the formation of nodes of Ranvier. Neuron 65:490–502

Fontana X, Hristova M, Da Costa C, Patodia S, Thei L, Makwana M et al (2012) c-Jun in Schwann cells promotes axonal regeneration and motoneuron survival via paracrine signaling. J Cell Biol 198:127–141

Fricker FR, Lago N, Balarajah S, Tsantoulas C, Tanna S, Zhu N et al (2011) Axonally derived neuregulin-1 is required for remyelination and regeneration after nerve injury in adulthood. J Neurosci 31:3225–3233

Jessen KR, Mirsky R (2005) The origin and development of glial cells in peripheral nerves. Nat Rev Neurosci 6:671–682

Jessen KR, Mirsky R (2008) Negative regulation of myelination: relevance for development, injury, and demyelinating disease. Glia 56:1552–1565

Liang C, Tao Y, Shen C, Tan Z, Xiong WC, Mei L (2012) Erbin is required for myelination in regenerated axons after injury. J Neurosci 32:15169–15180

Maurel P, Einheber S, Galinska J, Thaker P, Lam I, Rubin MB et al (2007) Nectin-like proteins mediate axon Schwann cell interactions along the internode and are essential for myelination. J Cell Biol 178:861–874

Nakata M, Baba H, Kanai K, Hoshi T, Sawai S, Hattori T et al (2008) Changes in Na^+ channel expression and nodal persistent Na^+ currents associated with peripheral nerve regeneration in mice. Muscle Nerve 37:721–730

Nave KA, Salzer JL (2006) Axonal regulation of myelination by neuregulin 1. Curr Opin Neurobiol 16:492–500

Pereira JA, Lebrun-Julien F, Suter U (2012) Molecular mechanisms regulating myelination in the peripheral nervous system. Trends Neurosci 35:123–134

Poliak S, Peles E (2003) The local differentiation of myelinated axons at nodes of Ranvier. Nat Rev Neurosci 4:968–980

Raivich G, Makwana M (2007) The making of successful axonal regeneration: genes, molecules and signal transduction pathways. Brain Res Rev 53:287–311

Rotshenker S (2011) Wallerian degeneration: the innate-immune response to traumatic nerve injury. J Neuroinflammation 8:109

Schäfer M, Fruttiger M, Montag D, Schachner M, Martini R (1996) Disruption of the gene for the myelin-associated glycoprotein improves axonal regrowth along myelin in C57BL/Wlds mice. Neuron 16:1107–1113

Shen YJ, DeBellard ME, Salzer JL, Roder J, Filbin MT (1998) Myelin-associated glycoprotein in myelin and expressed by Schwann cells inhibits axonal regeneration and branching. Mol Cell Neurosci 12:79–91

Sherman DL, Brophy PJ (2005) Mechanisms of axon ensheathment and myelin growth. Nat Rev Neurosci 6:683–690

Spiegel I, Adamsky K, Eshed Y, Milo R, Sabanay H, Sarig-Nadir O et al (2007) A central role for Necl4 (SynCAM4) in Schwann cell-axon interaction and myelination. Nat Neurosci 10:861–869

Stassart RM, Fledrich R, Velanac V, Brinkmann BG, Schwab MH, Meijer D et al (2013) A role for Schwann cell-derived neuregulin-1 in remyelination. Nat Neurosci 16:48–54

Susuki K, Rasband MN (2008) Molecular mechanisms of node of Ranvier formation. Curr Opin Cell Biol 20:616–623

Susuki K, Raphael AR, Ogawa Y, Stankewich MC, Peles E, Talbot WS et al (2011) Schwann cell spectrins modulate peripheral nerve myelination. Proc Natl Acad Sci USA 108:8009–8014

Taveggia C, Feltri ML, Wrabetz L (2010) Signals to promote myelin formation and repair. Nat Rev Neurol 6:276–287

Chapter 6
Charcot-Marie-Tooth Disease

Toru Ogata

Abstract Charcot-Marie-Tooth (CMT) disease is the most common disorder in hereditary peripheral nerve neuropathy. Depending on the clinical manifestation, the disease is divided into demyelinating type (CMT1, with autosomal dominant inheritance, and CMT4, with autosomal recessive inheritance), axonal type (CMT2, including both autosomal dominant and autosomal recessive inheritance), and combined type (dominant-intermediate CMT). To date, more than 40 genes have been reported in this field and for some of the causative genes more than 100 mutation sites have been identified. The relationship between genotype and phenotype is variable, with different clinical manifestations resulting from the same mutated gene.

Although our understanding of the pathogenesis of CMT remains limited, accumulating knowledge about the genetic etiology of this disease has provided information about both physiological and pathological myelin formation. In demyelinating CMT, Schwann cell functions are primarily impaired. Various causative mutant proteins are perceived to result in impairment of fundamental cell function during myelination and maintenance; myelin production, degradation of myelin protein (physiological or excessive), and endocytosis. These findings suggest that myelination requires precise regulation of a large amount of myelin proteins. This understanding of the underlying molecular mechanisms is expected to contribute to development of novel therapies for this hereditary neuropathy. Therapeutic approaches are being attempted in which the physiological function of Schwann cells is supported, impaired cell function is modified, and gene expressions are modified.

T. Ogata (✉)
Department of Rehabilitation for Movement Functions, Research Institute,
National Rehabilitation Center for Persons with Disabilities, 4-1 Namiki,
Tokorozawa, Saitama 359-8555, Japan
e-mail: ogata-toru@rehab.go.jp

K. Sango and J. Yamauchi (eds.), *Schwann Cell Development and Pathology*,
DOI 10.1007/978-4-431-54764-8_6, © Springer Japan 2014

Keywords Endoplasmic reticulum • Endosome • Genetic analyses • Intracellular trafficking • Lysosome • Protein misfolding • Rodent models • Ubiquitin–proteasome pathway

6.1 Introduction

Charcot-Marie-Tooth syndrome was originally described as a single disorder, in which patients manifest a progressive type of familial peroneal type of progressive muscular atrophy, by J.M. Charcot, P. Marie, and H.H. Tooth in 1886. Later, it was shown that there are several disorders involving peroneal muscle atrophy. The term hereditary motor and sensory neuropathy (HMSN) was introduced to encompass a broad group of inherited neuropathy syndromes that lack known metabolic abnormalities (Dyck and Lambert 1968). Now, inherited peripheral neuropathy encompasses HMSN (CMT), hereditary neuropathy with liability to pressure palsy (HNPP), hereditary motor neuropathy (HMN), and hereditary sensory and autonomic neuropathy (HSAN). Among those disorders, CMT disease is the most common clinical entity.

Classically, CMT is diagnosed and classified according to clinical symptoms, electrophysiological examination, and pathological features. Pathological findings enable the classification of the pathophysiology of CMT into either axonal damage dominant (CMT2), or demyelination dominant (CMT1 and CMT4) (Braathen 2012). Currently, CMT is distinguished according to the responsive genetic mutations, which have revealed the role of Schwann cells in the pathogenesis of the disease. In this chapter, we discuss the pathophysiology of CMT disease, focusing on Schwann cell functional deficits.

6.2 Clinical Features of Charcot-Marie-Tooth Disease

Patients with CMT begin to manifest symptoms in their teens to thirties, typically complaining about difficulties in locomotion. The distribution of peripheral nerve disturbance is indicative of polyneuropathy. Motor function is disturbed symmetrically and distally. Sensory disturbance indicates a distal glove and stocking pattern. Because of the conduction deficit in peripheral nerves in affected limbs, the deep tendon reflexes of the limbs are reduced.

6.3 Clinical Diagnosis

The clinical diagnosis of CMT starts with the history of symptoms, family history, and physical examination. When any type of hereditary peripheral neuropathy is suspected, electrodiagnostic tests are required. In examination of peripheral nerve conduction, a delayed conduction time implies the presence of demyelinating lesions and reduced compound muscle action potential (CMAP) implies axonal damages. The median motor nerve in the forearm is commonly used for examination of nerve conduction velocity (NCV), with a cutoff value of 38 m/s for distinguishing demyelinating types and axonal types (Dyck and Lambert 1968). When the affected patient's family shows variable NCVs ranging from 10 to 45 m/s, with dominant inheritance, the condition is defined as dominant-intermediate CMT (DI-CMT) (Davis et al. 1978). Together with the family history, autosomal dominant cases with demyelinating findings on a nerve conduction study are diagnosed as CMT1, whereas those with an autosomal recessive family history are diagnosed as CMT4. When males in the patients' family have severe clinical symptoms, this implies that the genetic abnormality is linked to the X chromosome, and the condition is termed CMTX. On the other hand, CMT2, an axonal type, comprises both autosomal dominant and recessive family histories. When these diagnostic procedures are not conclusive, further analysis, such as nerve biopsy or genetic investigation, should be performed.

Even after all available examinations have been performed, discrimination between axonal types and demyelinating types is not always clear. Axon–Schwann cell interaction is thought to be important, not only in myelin formation but also in maintenance of both cell types. Therefore, severe demyelinating pathology can lead to axonal damage and vice versa, leading to a mixture of both demyelinating lesions and axonal degeneration.

As for CMT1, the most common category, the typical histological findings comprise hypertrophic nerves with an onion bulb shape. Several clinical symptomatic entities are considered CMT-related diseases. Dejerine–Sottas disease (DSD, also called CMT3) is a clinical diagnosis in which patients show a severe and disabling neuropathy in infancy, starting at 3 years of age (Plante-Bordeneuve and Said 2002). Historically, DSD was first recognized as autosomal recessive inheritance, but later an autosomal dominant type was also reported. Congenital hypomyelination (CH or CHM) is also a clinical subtype of HMSN, in which patients have a developmental failure in myelination, manifesting delays in walking and swallowing, or even respiratory difficulties, as early as the first year of life. It is sometimes difficult to distinguish between DSD and CH, because both show markedly slow NCV in physiological examination and similar demyelinating histology in their sural nerve biopsies (Phillips et al. 1999).

Genetic testing is now performed for clinical purposes as well as for research. The precise genetic diagnosis may provide more information about the prognosis of each patient. Moreover, in the attempt to develop novel therapies for CMT, it is important to obtain a homogeneous population for clinical studies. In general, genetic analyses are useful for distinguishing subcategories of CMT1 (A–E) or CMT4 (A–J). However, it should be noted that the genetic background of CMT is not yet fully understood. Although more than 40 genes have been reported to be involved in CMT patients, mutations in those genes account for the disease in only approximately 70 % of CMT patients (Murphy et al. 2012). Therefore, a negative finding on genetic examination cannot exclude the possibility of CMT disease.

Table 6.1 Genes reported in Charcot-Marie-Tooth (CMT) diseases

Subtype	Inherent type	Causative genes	Molecular functions
CMT1A	AD	PMP22	Myelin structure
CMT1B	AD	MPZ (P0)	Myelin structure
CMT1C	AD	SIMPLE	Lysosomal function
CMT1D	AD	EGR2	Transcription factor
CMT1E	AD	NEFL	Cytoskeleton
CMT4A	AR	GDAP1	Mitochondrial function
CMT4B1	AR	MTMR2	Phosphatase in endosomal pathway
CMT4B2	AR	MTMR13 (SBF2)	Phosphatase in endosomal pathway
CMT4C	AR	SH3TC2	Endocytic recycling
CMT4D	AR	NDRG1	Trafficking
CMT4E	AR	EGR2	Transcription factor
CMT4F	AR	PRX	Interaction with extracellular matrix
CMT4G	AR	10q23	Unknown
CMT4H	AR	FGD4 (Frabin)	Activator of Cdc42, endosomal function
CMT4J	AR	FIG4	Phosphatase in endosomal pathway
CMTX1	X-linked	Cx32	Myelin structure
DI-CMTB	AD	DNM2	Endocytosis

AD autosomal dominant, *AR* autosomal recessive

In addition, several genes in CMT have been reported to be involved in more than two types of this disease. For example, mutations in *NEFL* have been reported in patients with CMT1E and CMT2E. Additionally, *EGR2* is known to be involved in CMT1D and CMT4E. Thus, genetic information has brought greater insights into this disorder; however, much work remains to be done to elucidate the link between genetic changes and disease manifestation.

6.4 Genes Reported in CMT

In demyelinating CMT disease, that is CMT1, CMT4, and CMT1X, 24 genes have been listed in the database: http://www.molgen.ua.ac.be/CMTMutations/Home/IPN.cfm (accessed in June 2013). Table 6.1 shows information about the genes reported in each of the subclasses of CMT1, CMT4, and CMT1X. In several subclasses, such as CMT4G, the responsible genes have not yet been identified and only limited information from linkage analyses is available.

6.5 Prevalence of CMT Diseases

Several studies have revealed the prevalence of CMT disease among inherited peripheral neuropathies. Additionally, recent genetic testing has facilitated data gathering on the frequency of mutations found in CMT disease (Braathen 2012;

Murphy et al. 2012). Although there are differences in the populations studied in each report, there are some consistent findings. CMT is the most common disease in inherited peripheral neuropathy. The prevalence of CMT1 is equal to or slightly higher than that of CMT2, whereas these two categories account for 70 % of all CMT disease. As for genetic testing results, abnormalities in four genes, PMP22, MPZ, Cx32, and MFN2 (CMT2), explain 70–90 % of all CMT cases (Saporta et al. 2011). Among them, duplication of PMP22 (CMT1A) is by far more common than other mutations, three- to fivefold more common than mutation of Cx32, the second most common genetic abnormality found in the CMT population. On the other hand, CMT4, a recessively inherited type of demyelinating CMT, is a rare form of CMT, and CMT4 as a whole accounts for less than 10 % of CMT diseases in European countries. However, the prevalence of CMT4 varies depending on area. In Mediterranean countries, where consanguineous marriages are prevalent, autosomal recessive inheritance accounts for 30–50 % of CMT cases (Dubourg et al. 2006). Within CMT4, CMT4A is the most common, and only a very limited number of families have been reported in some CMT4 subclasses, such as CMT4J.

6.6 Pathogenesis of CMT

The identification of specific gene mutations in CMT disease has yielded much insight into the mechanisms underlying CMT pathogenesis and Schwann cell functions during demyelination (Berger et al. 2006b). In some cases, the findings in human genetic testing were used to establish genetic mutant mice. Such animal models allow the use of various experimental methods to reveal molecular functions in the disease. It should be noted that although the causal link between specific gene mutations and CMT pathogenesis is quite clear, current knowledge is insufficient to explain clinical symptoms comprehensively, such as the age of onset and disease severity in each cases.

Hereafter, pathogenesis of CMT is discussed based on causal mutations in genes functioning in Schwann cells.

6.6.1 PMP22

Peripheral myelin protein 22 (PMP22) is a component of compact myelin that makes up 5 % of the peripheral myelin protein. *PMP22* is located on chromosome 17, which is the most common causative locus for CMT1 (autosomal dominant inheritance). This subgroup, termed CMT1A, is caused by several types of mutations in *PMP22*, via various mechanisms, resulting in diverse clinical manifestations.

6.6.1.1 Duplication

A large group of CMT1A patients has a 1.4-Mb chromosome 17 duplication, although some patients have been reported to have smaller duplications (Raeymaekers et al. 1989; Vance et al. 1989). Several animal studies have reported that rodent models overexpressing *PMP22* showed demyelination, the severity of which is related to the level of *PMP22* expression (Magyar et al. 1996; Sereda et al. 1996). Another transgenic mouse study reported that the demyelination in adult mice is reversible when excessive *PMP22* expression is switched off (Perea et al. 2001). Therefore, the gene dosage of *PMP22* is critical in the pathogenesis of CMT phenotype.

Although the precise mechanism by which excessive PMP22 protein leads to autosomal dominant demyelination is not yet fully clear, two possibilities have been proposed. First, PMP22 interacts with other myelin proteins, such as P0/MPZ, to form the proper myelin structure (Suter and Snipes 1995). Excess PMP22 may disturb the balance among myelin proteins, leading to destruction of myelin. The other possibility is that excessive PMP22 protein would be degraded by ubiquitination and proteasome pathways, which also maintain the appropriate expression level of PMP22 in healthy, myelinating Schwann cells. When the proteasome system is overwhelmed, the remaining PMP22 protein forms aggregates, which, in turn, exert toxic effects on Schwann cells (Fortun et al. 2005).

6.6.1.2 Point Mutation

Point mutations in PMP22 also cause autosomal dominant CMT, such as CMT1A. Among numerous mutation sites reported, the Leu16Pro mutation is common and is known as the Trembler-J mutation, which is the mutation that occurs in the spontaneous mutant mouse line (Suter et al. 1992a). Another mutation, Gly150Asp, Trembler mutation in mice, has been reported in DSD patients (Suter et al. 1992b; Ionasescu et al. 1997). Besides these missense mutations, other types of mutations, such as nonsense, frameshift, and splice site mutations, are rare and have been reported only in HNPP (Zephir et al. 2005). Because the majority of point mutations result in dominant inheritance of the disease, the mutated gene products likely cause a toxic gain of function.

At present, mutant PMP22 is thought to have a deficit in trafficking to the appropriate position in the plasma membrane. Some mutated proteins, such as Trembler mutation PMP22, can form heterodimers with the wild-type PMP22 protein and interfere with the trafficking of the normal protein (Tobler et al. 1999). Such unsorted proteins may form aggregates, as with duplication of *PMP22* gene, leading to overloading of the ubiquitin–proteasome pathway.

Alternatively, the mutated gene may produce unfolded or misfolded proteins. Such proteins, when arriving at the endoplasmic reticulum (ER), induce unfolded-protein reaction (UPR) (D'Urso et al. 1998). When excessive UPR occurs within the ER, it brings about a state of "ER stress," initiating apoptosis.

6.6.2 P0/MPZ

Myelin protein zero (P0/MPZ) is one of the myelin structural proteins and is also a member of the immunoglobulin superfamily. It is the most abundant protein constituting myelin (50 % of peripheral myelin protein) and also forms complexes with other myelin proteins such as PMP22. The corresponding gene, *MPZ*, is a causative gene for CMT1B, and more than 100 mutations, most of which are missense mutations, have been reported in this gene (Shy et al. 2004). These mutations predominantly occur in the extracellular domain, where extensive posttranslational modifications are added during transit through the ER and Golgi apparatus. The altered structure of MPZ induces destruction of the myelin structure, leading to early-onset demyelinating neuropathy (Shy 2006). It has also been suggested that protein misfolding and/or inadequate intracellular trafficking may induce ER stress and cell damage (Khajavi et al. 2005). From a clinical perspective, a distinct subgroup of MPZ mutations shows late-onset neuropathy with both axonal degeneration and mild demyelination. The pathogenesis of this subgroup may differ from the early-onset type and may include altered Schwann cell–axon interaction (Shy 2006).

It should be noted that there are similarities in clinical manifestations between *PMP22* and *MPZ* mutations (D'Urso et al. 1999). This similarity could be explained by the fact that both proteins play a similar physiological function in maintaining myelin structure and forming heterocomplexes to some extent. Furthermore, misfolding of these proteins has a dominant toxic effect on intracellular systems, such as the ER or ubiquitin–proteasome pathway.

6.6.3 SIMPLE

The small integral membrane protein of lysosome/late endosome (SIMPLE) is encoded on chromosome 16p13.1, a locus that is linked to CMT1C. Physiologically, SIMPLE has been reported to be involved in the ubiquitin–proteasome pathway. Although the physiological functions of the protein are not yet fully understood, SIMPLE interacts with the E3 ubiquitin ligase, NEDD4, mediating degradation of membrane proteins via the lysosomes (Shirk et al. 2005). Therefore, dysfunction of SIMPLE may lead to failure in turnover of myelin proteins, such as PMP22. Additionally, mutated SIMPLE itself participates in aggregate formation, which, in turn, places a load on the proteasome pathway. To date, six SIMPLE mutations have been reported, five of which have been reported in autosomal dominant CMT1C patients and one in autosomal dominant axonal type CMT (CMT2) (Houlden and Reilly 2006). Because the ubiquitin–proteasome pathway is involved in the pathogenesis of mutated myelin structural proteins (PMP22 and MPZ), the mechanisms by which SIMPLE mutations cause CMT disease may overlap with that of mutations involved in CMT1A and -1B. There has also been a report of a patient with both a SIMPLE mutation and a duplication of PMP22 who manifested with early-onset neuropathy, whereas the parents, with single mutations, had minimal symptoms (Meggouh et al. 2005).

6.6.4 *EGR2*

Early growth response 2 (EGR2) is a zinc finger transcription factor with the mouse orthologue Krox20. Krox20 plays a pivotal role in myelin formation by Schwann cells, by activating transcription of various myelin proteins, such as MPZ and myelin basic protein. Although Krox20-knockout mice show myelin formation failure, as expected, the heterozygous mice with half the normal dosage of Krox20 are phenotypically normal (Schneider-Maunoury et al. 1993). This, together with the fact that EGR2 mutations are mainly linked to a dominantly inherited disease, CMT1D, indicates that the pathogenesis of EGR2 mutations in CMT is not caused by the absence of this transcriptional factor but rather its dominant-negative function. Among nine mutation sites reported in *EGR*, all CMT1D patients have mutations in the zinc finger domain, which usually functions as a DNA-binding domain (Houlden and Reilly 2006). It has been reported that the presence of dominant mutations in Krox20 reduces transcription of *MPZ*. More specifically, dominant mutant Krox20 interferes with DNA binding of Sox10, another crucial transcription factor that binds adjacent to Krox20 in the *MPZ* regulatory region (LeBlanc et al. 2007). To date, it is not yet clear whether the same dominant-negative effects take place in the regulatory regions of other molecules regulated by Krox20.

There are also cases of recessive inheritance caused by mutations in *EGR2*, specifically in congenital hypomyelinating neuropathy (CMT4E). A mutation was found in the R1 domain (Ile268Asn), where NAB co-repressors bind. A deficit in the transcriptional complex formed between EGR2 (Krox20) and NAB proteins (NAB1 and NAB2) is thought to be responsible for failure in promoting myelin protein transcription, which leads to Schwann cell dysfunction in myelin formation (Le et al. 2005). Consistent with this proposal, NAB1 and NAB2 double-knockout mice present with severe congenital hypomyelination.

6.6.5 *NEFL*

Neurofilament protein, light filament (NEFL), is an intermediate neurofilament found in axons. It is the smallest member of three neurofilaments, known as NEFL (68 kDa), NF medium (125 kDa), and NF heavy (200 kDa). As expected, mutations in *NEFL* result in axonal CMT, such as CMT2E. Such mutations are also found in autosomal dominant CMT1 patients, such as CMT1F. It seems that the dominant pathogenic effect of mutant NEFL affects axons but not Schwann cells. Because NEFL plays a part in maintaining neurofilament assembly, point mutations cause dysfunction in axonal transport and mitochondria localization (Jordanova et al. 2003). These perturbations in axons may interfere with physiological axon–Schwann cell interaction, resulting in a demyelinating phenotype.

6.6.6 Connexin 32

In contrast to PMP22 and MPZ, which are localized in compact myelin, connexin 32 (Cx32/GJB1) is present in paranodal loops, internodal zones, and Schmidt–Lanterman incisures. Cx32/GJB1 is a member of gap junction-forming proteins and mediates the exchange of ions and metabolites. In the case of myelin-forming Schwann cells, the molecule connects different layers of myelin, allowing ions to move between the innermost and outermost myelin (Balice-Gordon et al. 1998). Although Cx32/GJB1-deficient mice can form functional gap junctions, the mice show demyelinating lesions and axonal loss, consistent with the symptoms found in CMTX1 (Scherer et al. 1998). Such phenotypes can be rescued by introducing Cx32/GJB1 expression specifically in Schwann cells, indicating that the pathogenesis is Schwann cell dependent and axonal loss is a secondary effect of Schwann cell dysfunction (Scherer et al. 2005).

More than 400 mutations related to Cx32/GJB1 have been reported in CMTX1 patients. These mutations are distributed, throughout all regions of the molecule, include nonsense, frameshift, and deletion mutations (Kleopa et al. 2012). As the background of various mutations shows a relatively similar degree of clinical symptoms, the pathomechanism of CMTX1 is caused by the loss of function of Cx32/GJB1. Indeed, several mutant proteins have been shown to form nonfunctional or abnormally functioning gap junction formation. Abnormal trafficking is also observed when mutations occur in the C-terminal region of the protein. Such abnormalities in trafficking lead to both changes in channel properties and accumulation of proteins in the ER and Golgi. Unlike PMP22 or MPZ, accumulation of Cx32/GJB1 has not been shown to overwhelm degradation pathways. Therefore, most of the mutations cause loss-of-function of proper channel function in Schwann cells. In rare cases, mutant proteins, such as those involving S85C and F235C, exert gain-of-function effects, by forming dysfunctional channels, resulting in female CTMX1 cases (Liang et al. 2005).

6.6.7 GDAP1

Mutations in the gene encoding ganglioside-induced differentiation-associated protein 1 (*GDAP1*) is most frequently found in recessively inherited CMT, named CMT4A, while mutations in the same gene have also been reported in other types of CMT (recessive-intermediate CMT A, CMT2K) (Cuesta et al. 2002). GDAP1 is located in the outer mitochondrial membrane, where it regulates mitochondrial function. More than 40 mutations in this gene have been reported, causing truncation of functional domains or mutations within the GST domains (Cassereau et al. 2011). Overexpression of wild-type *GDAP1*, but not *GDAP1* carrying a recessive mutation, increases the total cellular level of the antioxidant glutathione and increases the mitochondrial membrane potential (Noack et al. 2012). These findings suggest that loss of GDAP1 function leads to overproduction of reactive oxygen species (ROS) and activation of the related pathways, including apoptosis.

6.6.8 *MTMR2 and MTMR13*

Myotubularin-related protein (MTMR) constitutes a large family of phosphoinositide lipid 3-phosphatases that regulate the endosomal pathway. In CMT disease, mutations in the corresponding genes are found in autosomal recessive demyelinating cases, named CMT4B1 for *MTMR2* mutations and CMT4B2 for *MTMR13* mutations. MTMR2 and MTMR13 form heterocomplexes in which MTMR13 exerts a regulatory function on MTMR2 (Berger et al. 2006a). The synthesis of PI(3)P in the early endosome and PI(3,5)P2 in the late endosome is responsible for the formation of the endosome and associated lysosomal activities. MTMR family molecules are phosphatases for these phosphoinositides and are expected to function as regulators of the endosome system (Ng et al. 2013). A recent study showed that MTMR2 binds to various molecules, such as Dlg1, thereby regulating protein trafficking and membrane addition (Bolis et al. 2009). In spite of the fact that MTMR2 is not only expressed in Schwann cells, but also in neurons, Schwann cell-specific MTMR2 ablation in mutant mice confirmed that lack of MTMR2 in Schwann cells alone is sufficient to cause the demyelinating disease (Bolis et al. 2005). MTMR2-deficient mice are characterized by excessive redundant myelin, known as myelin outfoldings in CMT4B patients. Because the interaction of MTMR2 with Dlg1 negatively regulates membrane formation, a lack of MTMR2 induces excessive amounts of membrane.

6.6.9 *SH3TC2/KIAA1985*

The locus for CMT4C was assigned to chromosome 5q23–33 and mutations were identified in the SH3 domain and tetratricopeptide repeat domain 2 of *SH3TC2*, of which the product is expressed in the plasma membrane and perinuclear endocytic recycling compartment (Arnaud et al. 2009). Recently, molecular interaction was found between SH3TC2 and Rab11, a pivotal regulatory molecule involved in recycling of internalized membranes and receptors back to the cell surface in the endosome trafficking pathway. Mutations in *SH3TC2* disturb SH3TC2/Rab11 complex formation and impair endocytic recycling (Stendel et al. 2010). Although the dominant-negative form of Rab11 has been reported to impair myelin formation, it remains unknown which constituent of myelin protein is specifically transported by SH3TC2/Rab11-dependent trafficking.

6.6.10 *NDRG1*

N-myc downstream-regulated gene 1 (*NDRG1*) is ubiquitously expressed and plays a role in growth arrest and cell differentiation (Melotte et al. 2010). A lack of NDRG1 impairs trafficking in the cytoplasm and nucleus, similar to that seen with *SH3TC2* (CMT4C) mutation (Kalaydjieva et al. 2000; King et al. 2011). The mutation in *NDRG1* was initially reported in Lom-type HMNS (HMNSL), and the clinical subclass is named CMT4D. In the clinical specimen, demyelination occurred together with severe axonal loss, suggesting impairment in axon–glial interaction.

6.6.11 PRX

Periaxin (PRX) is a member of the PDZ-domain proteins, having two isoforms, L-periaxin and S-periaxin. The PDZ domain facilitates interaction with other molecules, forming dystroglycan–dystrophin-related protein 2 (DRP2) complexes at the cell membrane of myelinating Schwann cells. This complex links the Schwann cell cytoskeleton to the extracellular matrix, and laminin in particular, and plays a crucial role in maintaining myelin structure (Sherman et al. 2001). Sixteen nonsense or frameshift mutations have been reported in *PRX* among autosomal recessive CMT patients (CMT4F) and Dejerine–Sottas disease patients. Because the phenotype of PRX-deficient mice recaptures the clinical symptoms of human patients, the demyelinating lesion is likely a loss of function of PRX. In some clinical cases, a truncated form of PRX is found complexed with DRP2, suggesting that pathogenesis of the recessive phenotype is independent of the interaction between PRX and DRP2 (Takashima et al. 2002). There is also a possibility that the truncated form of PRX is resistant to degradation, which produces gain-of-function effects on myelin stability.

6.6.12 Frabin/FGD4

Frabin, known as a causative gene for CMT4H, is a member of the GDP/GTP nucleotide exchange factors (GEF) and is a specific activator of Cdc42, a member of the Rho GTPase family, which regulate assembly of the actin cytoskeleton and microtubules (Delague et al. 2007; Horn et al. 2012). Cdc42 is reported to be involved in cell migration, polarization, division, and membrane trafficking. Besides a Dbl homology (DH) domain, the region responsible for regulating Cdc42 function, Frabin also contains other domains, such as a pleckstrin homology (PH) domain and cysteine-rich FYVE domain. These domains are known to be necessary for activation of Cdc42. The presence of PH and FYVE domains also is important for activating c-Jun N-terminal kinase (JNK), as well as for facilitating the binding of Frabin to phosphoinositide phosphates (PIPs), substrates of the myotubularin-related proteins MTMR2 and MTMR13, which are phosphatases for PIPs (Stendel et al. 2007). As mutations in *MTMR2* and *MTMR13* impair endosomal regulation of myelin proteins, leading to CMT4B1 and CMTB2, respectively, association between Frabin and MTMRs may characterize the importance of Frabin in membrane transport in myelinating Schwann cells. It is noteworthy that pathological findings between CMT4B and CMT4H are similar; that is, both conditions are characterized by myelin outfoldings, irregular folding, and redundant loops of myelin (Tazir et al. 2013).

6.6.13 *FIG4*

FIG4 is a phosphatase acting on the 5-phosphate from phosphoinositide PI $(3,5)P_2$ on the surface of vesicles in the endosome/lysosome pathway. FIG4 acts on the same target as MTMR2, which is causative for CMT4B, but removes a different phosphate. Mutations in *FIG4* have been reported in a severely recessive subgroup of CMT, CMT4J (Chow et al. 2007). The majority of the patients are compound heterozygotes carrying the missense allele I41T, in combination with a null allele of *FIG4*. The I41T mutant can exert partial functions of FIG4 but is unstable because of impairment of its interaction with the scaffold protein, VAC14. The loss of function of FIG4 results in accumulation of endosomal/lysosomal vesicles, affecting proper myelin maintenance (Lenk et al. 2011).

6.6.14 *DNM2*

Dynamin 2 (DNM2) is a member of the large GTPase family, regulating membrane trafficking from the trans-Golgi network, actin cytoskeletal dynamics, and membrane fusion (Praefcke and McMahon 2004). Dynamin is recruited to the site of endocytosis and its GTPase activity provides energy for the endocytosis processes. *DNM2* is expressed ubiquitously and is reportedly linked to a dominant-intermediate CMT (DI-CMTB). Most of the reported mutation sites among DI-CMTB patients are located around the PH domain that mediates interaction with phosphoinositides, recruiting DNM2 to the vesicles. The mutated proteins have reduced capacity to bind to vesicles, resulting in impairment of endocytosis (Zuchner et al. 2005).

6.7 General Aspects of CMT Pathogenesis

As listed here, there are many genes and mutations reported to be involved in the pathogenesis of demyelinating CMT diseases. In each subgroup, further detail in our understanding of the function of the affected genes in physiology and pathology is required. However, to develop therapeutic strategies for CMT diseases, it would be advantageous to perceive the pathogenesis of the disease from the point of view of generalized pathways involved in demyelinating CMT.

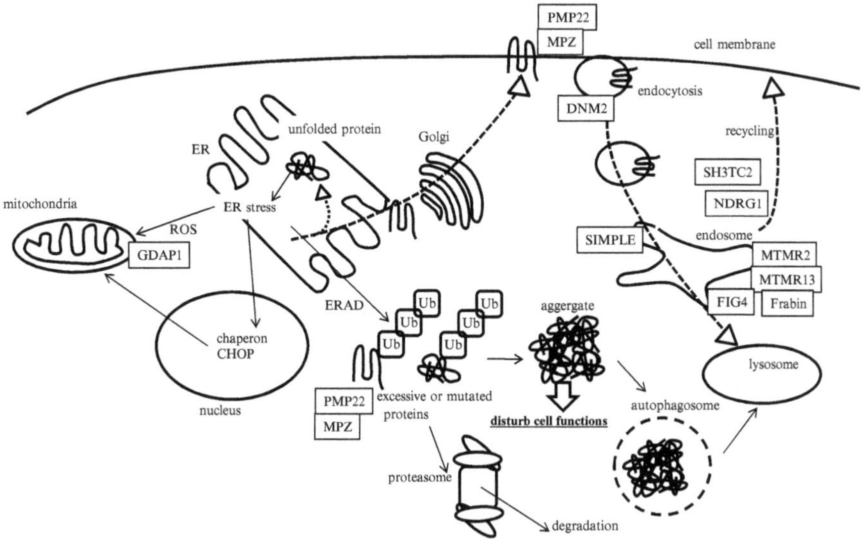

Fig. 6.1 Pathways for protein degradation in Schwann cells. The production of myelin proteins is tightly regulated with several pathways to degrade undesired proteins. Many causative genes in Charcot-Marie-Tooth (CMT) diseases are related to these pathways, indicating their importance. *ER* endoplasmic reticulum, *ERAD* ER-associated degradation, *Ub* ubiquitination, *ROS* reactive oxygen species

In a simplified summary, the mechanisms underlying demyelinating CMT involve disruption of the membrane structure or accumulation of abnormal or overexpressed proteins. Furthermore, disruption of the membrane can be caused by a lack of components, which usually is prominent in recessively inherited conditions, or the presence of abnormal (or excessive) proteins, by which the physiological molecular interaction in the membrane is disturbed in a dominantly inherited manner.

Accumulation of myelin or membrane proteins may exert harmful effects upon Schwann cells through various pathways. To understand those pathways, it is important to understand the physiological protein quality control mechanisms within Schwann cells. It has been reported that up to 80 % of the newly synthesized PMP22 is degraded within 30 min via the proteasome, because of misfolding of the protein or excessive production (Ryan et al. 2002). Such high turnover maintenance of myelin protein is one of the reasons that dysfunction of the degradative pathways can lead to accumulation of the protein. During the process of myelination and remyelination, transactivation of myelin proteins is upregulated, so that heterogeneous mutations can result in excessive levels of myelin proteins, which cannot be integrated into the myelin membrane and are destined to be degraded. Figure 6.1 summarizes the intracellular systems that handle misfolded or excessive proteins.

6.7.1 UPR at ER

Misfolded myelin proteins are retained in the ER and their accumulation activates the unfolded protein response (UPR), characterized by activation of Ire-1, Perk, and ATF6 pathways (Ron and Walter 2007). As an adaptive response, UPR induces specific chaperone expression and at the same time reduces translation, to prevent further accumulation of the unfolded proteins. Transgenic mice models overexpressing wild or mutant forms of MPZ showed that although the S63del mutant is retained in the ER and induces the UPR response, normal MPZ, another MPZ mutant (S63C), and the Trembler-J mutant of PMP22 do not activate the UPR (Pennuto et al. 2008). Although the UPR itself is a physiological component of cell maintenance, severe ER stress can also induce the pro-apoptotic molecule, CHOP. In general, induction of CHOP expression after ER stress triggers the cell death program to eliminate the affected cells. The involvement of CHOP in the pathogenesis of the MPZ S63del is proven by the fact that ablation of Chop from S63del-transgenic mice results in a reduction of demyelinating lesions. Interestingly, the number of Schwann cells is not altered significantly by the absence of CHOP; it is now understood that activation of the CHOP pathway induces dysfunction of Schwann cells, which is independent of cell death (Pennuto et al. 2008).

6.7.2 Ubiquitin–Proteasome Pathway

When misfolded proteins are not fully refolded in the ER, these proteins are released into the cytosol, where specific E3 ligases target the molecule and add several ubiquitin molecules. The poly-ubiquitinated protein is then recognized by proteasome systems, including the lysosome where the protein is finally degraded. In CMT disease, excessive PMP22, mutant PMP22, and mutant MPZ have been reported to be processed by the proteasome via the ER, whereas mutant SIMPLE (CMT1C) is degraded by the proteasome in ER-independent mechanisms (Lee et al. 2012). The function of proteasome is inhibited by oxidative stress, age, anti-cancer drugs, the presence of mutant PMP22 (Trembler-J mutant), or the presence of other protein aggregates. Impaired function of the proteasome leads to a further increase of cytoplasmic protein aggregates (Fortun et al. 2006). The presence of aggregates is thought to be a causative event, not only in CMT, but also in various degenerative diseases, such as Huntington disease. One of the possible mechanisms is that the aggregates entrap other molecules to form nonfunctional complexes. In the case of Trembler-J mice, MBP is mislocalized to PMP22 protein aggregates, reducing the proper sorting of MBP into the compact myelin (Fortun et al. 2005). Although protein aggregates exert cytotoxic effects in other degenerative diseases, whether myelin protein aggregates induce cell death in Schwann cells is not yet clear.

6.7.3 Aggresome–Autophagy Pathway

Another protein quality control system involved in degrading misfolded and aggregated proteins is the aggresome–autophagy pathway. Aggregated proteins that are not handled by the proteasome pathway are sequestered to the perinuclear aggresome through microtubule-dependent transport. The aggresomes are then surrounded by the autophagosome and degraded via autophagy. Because autophagy is an alternative way to handle abnormal protein aggregates, some attempts have been made to enhance autophagy activity in CMT model mice to reduce the amount of aggregates (Fortun et al. 2007).

6.7.4 Inflammation

Although the majority of studies of demyelinating CMT diseases focused on intrinsic pathological events in Schwann cells, the secondary involvement of extrinsic events may be crucial in determining disease progression and severity. The presence of low-grade inflammation is reported both in specimens of CMT patients and in mice models (Malandrini et al. 1999; Kohl et al. 2010). In the mouse model, the number of macrophages infiltrated into the peripheral nerves correlates with axonal damage. Among numerous inflammatory mediators, MCP-1/CCL2 is thought to be a crucial molecule in PMP22tg mice, because ablation of the gene encoding this chemokine can prevent accumulation of macrophages (Kohl et al. 2010). The involvement of inflammation in CMT disease is also suggested from the study of crossbreeding of Cx32-deficient mice (a model of CMTX1) with colony-stimulating factor-1 (CSF-1)-deficient mice. The lack of CSF-1, a cytokine that recruits and activates macrophages, leads to reduction of demyelinating lesions in the mouse model (Groh et al. 2012). The source of these cytokines may not be restricted to Schwann cells, but may also involve other structural cells in the peripheral nerves, such as endoneural fibroblasts.

6.8 Therapeutic Approaches for CMT Diseases

It has been described that there is no established disease-modifying therapy for CMT disease, although supportive care, including use of orthotics and pain management, have been well developed. Nevertheless, the accumulation of knowledge about the cellular and molecular mechanisms of myelination and demyelination provides several possibilities to approach the disease (Herrmann 2008).

6.8.1 Ascorbic Acid

The necessity of ascorbic acid for myelin formation by Schwann cells has been well documented in Schwann cell–dorsal root ganglion co-culture studies. Ascorbic acid is required to form the basement membrane before the initiation of myelination. These in vitro studies provide the rationale for an animal study in which ascorbic acid was given to a mouse model of CMT1A (Passage et al. 2004). In these experiments, ascorbic acid reduced *PMP22* mRNA at the transcriptional level. Based on these preclinical data, two clinical trials of ascorbic acid in CMT1A patients are currently underway (Lewis et al. 2013; Pareyson et al. 2006).

6.8.2 Progesterone Antagonist

Progesterone has been shown to promote expression of myelin-related genes, such as *PMP22* and *MPZ*. Therefore, antagonizing its function may convey beneficial effects in the case of excessive myelin protein production, as occurs with duplication of *PMP22* (CMT1A). A progesterone receptor antagonist, onapristone, improved the pathological phenotype of CMT1A model mice when the chemical was applied in neonates (Sereda et al. 2003). However, the further clinical application of onapristone is not promising, as this drug is unsafe for use in humans.

6.8.3 Neurotrophin-3

Schwann cell myelination is influenced by various extrinsic molecules, including nerve growth factors. Among these, neurotrophin-3 (NT-3) is a promising molecule for promoting myelin formation. Elevated myelin formation has been shown in both in vitro and in vivo experiments with NT-3 treatment. A pilot study using NT-3 in CMT1A patients reported some beneficial effects of this treatment (Sahenk et al. 2005).

6.8.4 RNA and Gene-Based Therapy

It is assumed that the development of gene-based technologies will provide a better opportunity for establishing therapeutic approaches by directly targeting mRNA or genes. One of the most promising methods would be RNA interference methods. By using short interfering RNA, it is possible to downregulate the expression of undesired mRNA specifically (Bhindi et al. 2007). Therefore, the gain-of-function mutations in CMT, mainly CMT1, would be an ideal target for this approach.

However, because CMT disease is a congenital and lifelong disorder, undesired gene expression should be regulated throughout life. Considering that the lesion occurs within the peripheral nerve, which is more difficult to access as compared to muscles, these drug delivery issues should be resolved in future.

6.9 Conclusion

To understand the pathogenesis of CMT, a focus on Schwann cell functions has given us the opportunity to understand how myelination is regulated and maintained in peripheral nerves. The fact that several causative gene products in CMT are expressed ubiquitously, but result specifically in peripheral neuropathy, indicates that achieving proper localization of adequate amounts of myelin proteins requires precise performance of various cell functions. Therefore, Schwann cell myelination would be an informative experimental model for studying cell functions, such as membrane trafficking and degradation. With respect to future challenges for novel therapeutic approaches, it should be kept in mind that the majority of CMT patients bear mutations in *PMP22* or *MPZ* (CMT1A and -B), with a minor population bearing mutations in other genes. In the case of recessive CMT4, it is not realistic to organize a clinical study including more than 100 patients because the patient population for each subtype is too small. Therefore, basic research into CMT should explore mechanisms of general myelination pathways further to contribute to CMT-related diseases as a whole.

References

Arnaud E, Zenker J, de Preux Charles AS, Stendel C, Roos A, Medard JJ, Tricaud N, Kleine H, Luscher B, Weis J, Suter U, Senderek J, Chrast R (2009) SH3TC2/KIAA1985 protein is required for proper myelination and the integrity of the node of Ranvier in the peripheral nervous system. Proc Natl Acad Sci USA 106:17528–17533

Balice-Gordon RJ, Bone LJ, Scherer SS (1998) Functional gap junctions in the Schwann cell myelin sheath. J Cell Biol 142:1095–1104

Berger P, Berger I, Schaffitzel C, Tersar K, Volkmer B, Suter U (2006a) Multi-level regulation of myotubularin-related protein-2 phosphatase activity by myotubularin-related protein-13/set-binding factor-2. Hum Mol Genet 15:569–579

Berger P, Niemann A, Suter U (2006b) Schwann cells and the pathogenesis of inherited motor and sensory neuropathies (Charcot-Marie-Tooth disease). Glia 54:243–257

Bhindi R, Fahmy RG, Lowe HC, Chesterman CN, Dass CR, Cairns MJ, Saravolac EG, Sun LQ, Khachigian LM (2007) Brothers in arms: DNA enzymes, short interfering RNA, and the emerging wave of small-molecule nucleic acid-based gene-silencing strategies. Am J Pathol 171:1079–1088

Bolis A, Coviello S, Bussini S, Dina G, Pardini C, Previtali SC, Malaguti M, Morana P, Del Carro U, Feltri ML, Quattrini A, Wrabetz L, Bolino A (2005) Loss of Mtmr2 phosphatase in Schwann cells but not in motor neurons causes Charcot-Marie-Tooth type 4B1 neuropathy with myelin outfoldings. J Neurosci 25:8567–8577

Bolis A, Coviello S, Visigalli I, Taveggia C, Bachi A, Chishti AH, Hanada T, Quattrini A, Previtali SC, Biffi A, Bolino A (2009) Dlg1, Sec8, and Mtmr2 regulate membrane homeostasis in Schwann cell myelination. J Neurosci 29:8858–8870

Braathen GJ (2012) Genetic epidemiology of Charcot-Marie-Tooth disease. Acta Neurol Scand 126(Suppl 193):1–22

Cassereau J, Chevrollier A, Gueguen N, Desquiret V, Verny C, Nicolas G, Dubas F, Amati-Bonneau P, Reynier P, Bonneau D, Procaccio V (2011) Mitochondrial dysfunction and pathophysiology of Charcot-Marie-Tooth disease involving GDAP1 mutations. Exp Neurol 227:31–41

Chow CY, Zhang Y, Dowling JJ, Jin N, Adamska M, Shiga K, Szigeti K, Shy ME, Li J, Zhang X, Lupski JR, Weisman LS, Meisler MH (2007) Mutation of FIG4 causes neurodegeneration in the pale tremor mouse and patients with CMT4J. Nature (Lond) 448:68–72

Cuesta A, Pedrola L, Sevilla T, Garcia-Planells J, Chumillas MJ, Mayordomo F, Leguern E, Marin I, Vilchez JJ, Palau F (2002) The gene encoding ganglioside-induced differentiation-associated protein 1 is mutated in axonal Charcot-Marie-Tooth type 4A disease. Nat Genet 30:22–25

D'Urso D, Prior R, Greiner-Petter R, Gabreels-Festen AA, Muller HW (1998) Overloaded endoplasmic reticulum-Golgi compartments, a possible pathomechanism of peripheral neuropathies caused by mutations of the peripheral myelin protein PMP22. J Neurosci 18:731–740

D'Urso D, Ehrhardt P, Muller HW (1999) Peripheral myelin protein 22 and protein zero: a novel association in peripheral nervous system myelin. J Neurosci 19:3396–3403

Davis CJ, Bradley WG, Madrid R (1978) The peroneal muscular atrophy syndrome: clinical, genetic, electrophysiological and nerve biopsy studies. I. Clinical, genetic and electrophysiological findings and classification. J Hum Genet 26:311–349

Delague V, Jacquier A, Hamadouche T, Poitelon Y, Baudot C, Boccaccio I, Chouery E, Chaouch M, Kassouri N, Jabbour R, Grid D, Megarbane A, Haase G, Levy N (2007) Mutations in FGD4 encoding the Rho GDP/GTP exchange factor FRABIN cause autosomal recessive Charcot-Marie-Tooth type 4H. Am J Hum Genet 81:1–16

Dubourg O, Azzedine H, Verny C, Durosier G, Birouk N, Gouider R, Salih M, Bouhouche A, Thiam A, Grid D, Mayer M, Ruberg M, Tazir M, Brice A, Leguern E (2006) Autosomal-recessive forms of demyelinating Charcot-Marie-Tooth disease. Neuromolecular Med 8:75–86

Dyck PJ, Lambert EH (1968) Lower motor and primary sensory neuron diseases with peroneal muscular atrophy. I. Neurologic, genetic, and electrophysiologic findings in hereditary polyneuropathies. Arch Neurol 18:603–618

Fortun J, Li J, Go J, Fenstermaker A, Fletcher BS, Notterpek L (2005) Impaired proteasome activity and accumulation of ubiquitinated substrates in a hereditary neuropathy model. J Neurochem 92:1531–1541

Fortun J, Go JC, Li J, Amici SA, Dunn WA Jr, Notterpek L (2006) Alterations in degradative pathways and protein aggregation in a neuropathy model based on PMP22 overexpression. Neurobiol Dis 22:153–164

Fortun J, Verrier JD, Go JC, Madorsky I, Dunn WA, Notterpek L (2007) The formation of peripheral myelin protein 22 aggregates is hindered by the enhancement of autophagy and expression of cytoplasmic chaperones. Neurobiol Dis 25:252–265

Groh J, Weis J, Zieger H, Stanley ER, Heuer H, Martini R (2012) Colony-stimulating factor-1 mediates macrophage-related neural damage in a model for Charcot-Marie-Tooth disease type 1X. Brain 135:88–104

Herrmann DN (2008) Experimental therapeutics in hereditary neuropathies: the past, the present, and the future. Neurotherapeutics 5:507–515

Horn M, Baumann R, Pereira JA, Sidiropoulos PN, Somandin C, Welzl H, Stendel C, Luhmann T, Wessig C, Toyka KV, Relvas JB, Senderek J, Suter U (2012) Myelin is dependent on the Charcot-Marie-Tooth Type 4H disease culprit protein FRABIN/FGD4 in Schwann cells. Brain 135:3567–3583

Houlden H, Reilly MM (2006) Molecular genetics of autosomal-dominant demyelinating Charcot-Marie-Tooth disease. Neuromolecular Med 8:43–62

Ionasescu VV, Searby CC, Ionasescu R, Chatkupt S, Patel N, Koenigsberger R (1997) Dejerine-Sottas neuropathy in mother and son with same point mutation of PMP22 gene. Muscle Nerve 20:97–99

Jordanova A, De Jonghe P, Boerkoel CF, Takashima H, De Vriendt E, Ceuterick C, Martin JJ, Butler IJ, Mancias P, Papasozomenos S, Terespolsky D, Potocki L, Brown CW, Shy M, Rita DA, Tournev I, Kremensky I, Lupski JR, Timmerman V (2003) Mutations in the neurofilament light chain gene (NEFL) cause early onset severe Charcot-Marie-Tooth disease. Brain 126:590–597

Kalaydjieva L, Gresham D, Gooding R, Heather L, Baas F, De Jonge R, Blechschmidt K, Angelicheva D, Chandler D, Worsley P, Rosenthal A, King RH, Thomas PK (2000) N-myc downstream-regulated gene 1 is mutated in hereditary motor and sensory neuropathy-Lom. Am J Hum Genet 67:47–58

Khajavi M, Inoue K, Wiszniewski W, Ohyama T, Snipes GJ, Lupski JR (2005) Curcumin treatment abrogates endoplasmic reticulum retention and aggregation-induced apoptosis associated with neuropathy-causing myelin protein zero-truncating mutants. Am J Hum Genet 77:841–850

King RH, Chandler D, Lopaticki S, Huang D, Blake J, Muddle JR, Kilpatrick T, Nourallah M, Miyata T, Okuda T, Carter KW, Hunter M, Angelicheva D, Morahan G, Kalaydjieva L (2011) Ndrg1 in development and maintenance of the myelin sheath. Neurobiol Dis 42:368–380

Kleopa KA, Abrams CK, Scherer SS (2012) How do mutations in GJB1 cause X-linked Charcot-Marie-Tooth disease? Brain Res 1487:198–205

Kohl B, Fischer S, Groh J, Wessig C, Martini R (2010) MCP-1/CCL2 modifies axon properties in a PMP22-overexpressing mouse model for Charcot-Marie-tooth 1A neuropathy. Am J Pathol 176:1390–1399

Le N, Nagarajan R, Wang JY, Svaren J, Lapash C, Araki T, Schmidt RE, Milbrandt J (2005) Nab proteins are essential for peripheral nervous system myelination. Nat Neurosci 8:932–940

LeBlanc SE, Ward RM, Svaren J (2007) Neuropathy-associated Egr2 mutants disrupt cooperative activation of myelin protein zero by Egr2 and Sox10. Mol Cell Biol 27:3521–3529

Lee SM, Chin LS, Li L (2012) Protein misfolding and clearance in demyelinating peripheral neuropathies: therapeutic implications. Commun Integr Biol 5:107–110

Lenk GM, Ferguson CJ, Chow CY, Jin N, Jones JM, Grant AE, Zolov SN, Winters JJ, Giger RJ, Dowling JJ, Weisman LS, Meisler MH (2011) Pathogenic mechanism of the FIG4 mutation responsible for Charcot-Marie-Tooth disease CMT4J. PLoS Genet 7:e1002104

Lewis RA, McDermott MP, Herrmann DN, Hoke A, Clawson LL, Siskind C, Feely SM, Miller LJ, Barohn RJ, Smith P, Luebbe E, Wu X, Shy ME (2013) High-dosage ascorbic acid treatment in Charcot-Marie-Tooth disease type 1A: results of a randomized, double-masked, controlled trial. JAMA Neurol 70(8):981–987

Liang GS, de Miguel M, Gomez-Hernandez JM, Glass JD, Scherer SS, Mintz M, Barrio LC, Fischbeck KH (2005) Severe neuropathy with leaky connexin 32 hemichannels. Ann Neurol 57:749–754

Magyar JP, Martini R, Ruelicke T, Aguzzi A, Adlkofer K, Dembic Z, Zielasek J, Toyka KV, Suter U (1996) Impaired differentiation of Schwann cells in transgenic mice with increased PMP22 gene dosage. J Neurosci 16:5351–5360

Malandrini A, Villanova M, Dotti MT, Federico A (1999) Acute inflammatory neuropathy in Charcot-Marie-Tooth disease. Neurology 52:859–861

Meggouh F, de Visser M, Arts WF, De Coo RI, van Schaik IN, Baas F (2005) Early onset neuropathy in a compound form of Charcot-Marie-Tooth disease. Ann Neurol 57:589–591

Melotte V, Qu X, Ongenaert M, van Criekinge W, de Bruine AP, Baldwin HS, van Engeland M (2010) The N-myc downstream regulated gene (NDRG) family: diverse functions, multiple applications. FASEB J 24:4153–4166

Murphy SM, Laura M, Fawcett K, Pandraud A, Liu YT, Davidson GL, Rossor AM, Polke JM, Castleman V, Manji H, Lunn MP, Bull K, Ramdharry G, Davis M, Blake JC, Houlden H, Reilly MM (2012) Charcot-Marie-Tooth disease: frequency of genetic subtypes and guidelines for genetic testing. J Neurol Neurosurg Psychiatry 83:706–710

Ng AA, Logan AM, Schmidt EJ, Robinson FL (2013) The CMT4B disease-causing phosphatases Mtmr2 and Mtmr13 localize to the Schwann cell cytoplasm and endomembrane compartments, where they depend upon each other to achieve wild-type levels of protein expression. Hum Mol Genet 22:1493–1506

Noack R, Frede S, Albrecht P, Henke N, Pfeiffer A, Knoll K, Dehmel T, Meyer Zu Horste G, Stettner M, Kieseier BC, Summer H, Golz S, Kochanski A, Wiedau-Pazos M, Arnold S, Lewerenz J, Methner A (2012) Charcot-Marie-Tooth disease CMT4A: GDAP1 increases cellular glutathione and the mitochondrial membrane potential. Hum Mol Genet 21:150–162

Pareyson D, Schenone A, Fabrizi GM, Santoro L, Padua L, Quattrone A, Vita G, Gemignani F, Visioli F, Solari A (2006) A multicenter, randomized, double-blind, placebo-controlled trial of long-term ascorbic acid treatment in Charcot-Marie-Tooth disease type 1A (CMT-TRIAAL): the study protocol [EudraCT no.: 2006-000032-27]. Pharmacol Res 54:436–441

Passage E, Norreel JC, Noack-Fraissignes P, Sanguedolce V, Pizant J, Thirion X, Robaglia-Schlupp A, Pellissier JF, Fontes M (2004) Ascorbic acid treatment corrects the phenotype of a mouse model of Charcot-Marie-Tooth disease. Nat Med 10:396–401

Pennuto M, Tinelli E, Malaguti M, Del Carro U, D'Antonio M, Ron D, Quattrini A, Feltri ML, Wrabetz L (2008) Ablation of the UPR-mediator CHOP restores motor function and reduces demyelination in Charcot-Marie-Tooth 1B mice. Neuron 57:393–405

Perea J, Robertson A, Tolmachova T, Muddle J, King RH, Ponsford S, Thomas PK, Huxley C (2001) Induced myelination and demyelination in a conditional mouse model of Charcot-Marie-Tooth disease type 1A. Hum Mol Genet 10:1007–1018

Phillips JP, Warner LE, Lupski JR, Garg BP (1999) Congenital hypomyelinating neuropathy: two patients with long-term follow-up. Pediatr Neurol 20:226–232

Plante-Bordeneuve V, Said G (2002) Dejerine-Sottas disease and hereditary demyelinating polyneuropathy of infancy. Muscle Nerve 26:608–621

Praefcke GJ, McMahon HT (2004) The dynamin superfamily: universal membrane tubulation and fission molecules? Nat Rev Mol Cell Biol 5:133–147

Raeymaekers P, Timmerman V, De Jonghe P, Swerts L, Gheuens J, Martin JJ, Muylle L, De Winter G, Vandenberghe A, Van Broeckhoven C (1989) Localization of the mutation in an extended family with Charcot-Marie-Tooth neuropathy (HMSN I). Am J Hum Genet 45:953–958

Ron D, Walter P (2007) Signal integration in the endoplasmic reticulum unfolded protein response. Nat Rev Mol Cell Biol 8:519–529

Ryan MC, Shooter EM, Notterpek L (2002) Aggresome formation in neuropathy models based on peripheral myelin protein 22 mutations. Neurobiol Dis 10:109–118

Sahenk Z, Nagaraja HN, McCracken BS, King WM, Freimer ML, Cedarbaum JM, Mendell JR (2005) NT-3 promotes nerve regeneration and sensory improvement in CMT1A mouse models and in patients. Neurology 65:681–689

Saporta AS, Sottile SL, Miller LJ, Feely SM, Siskind CE, Shy ME (2011) Charcot-Marie-Tooth disease subtypes and genetic testing strategies. Ann Neurol 69:22–33

Scherer SS, Xu YT, Nelles E, Fischbeck K, Willecke K, Bone LJ (1998) Connexin32-null mice develop demyelinating peripheral neuropathy. Glia 24:8–20

Scherer SS, Xu YT, Messing A, Willecke K, Fischbeck KH, Jeng LJ (2005) Transgenic expression of human connexin32 in myelinating Schwann cells prevents demyelination in connexin32-null mice. J Neurosci 25:1550–1559

Schneider-Maunoury S, Topilko P, Seitandou T, Levi G, Cohen-Tannoudji M, Pournin S, Babinet C, Charnay P (1993) Disruption of Krox-20 results in alteration of rhombomeres 3 and 5 in the developing hindbrain. Cell 75:1199–1214

Sereda M, Griffiths I, Puhlhofer A, Stewart H, Rossner MJ, Zimmerman F, Magyar JP, Schneider A, Hund E, Meinck HM, Suter U, Nave KA (1996) A transgenic rat model of Charcot-Marie-Tooth disease. Neuron 16:1049–1060

Sereda MW, Meyer Zu Horste G, Suter U, Uzma N, Nave KA (2003) Therapeutic administration of progesterone antagonist in a model of Charcot-Marie-Tooth disease (CMT-1A). Nat Med 9:1533–1537

Sherman DL, Fabrizi C, Gillespie CS, Brophy PJ (2001) Specific disruption of a Schwann cell dystrophin-related protein complex in a demyelinating neuropathy. Neuron 30:677–687

Shirk AJ, Anderson SK, Hashemi SH, Chance PF, Bennett CL (2005) SIMPLE interacts with NEDD4 and TSG101: evidence for a role in lysosomal sorting and implications for Charcot-Marie-Tooth disease. J Neurosci Res 82:43–50

Shy ME (2006) Peripheral neuropathies caused by mutations in the myelin protein zero. J Neurol Sci 242:55–66

Shy ME, Jani A, Krajewski K, Grandis M, Lewis RA, Li J, Shy RR, Balsamo J, Lilien J, Garbern JY, Kamholz J (2004) Phenotypic clustering in MPZ mutations. Brain 127:371–384

Stendel C, Roos A, Deconinck T, Pereira J, Castagner F, Niemann A, Kirschner J, Korinthenberg R, Ketelsen UP, Battaloglu E, Parman Y, Nicholson G, Ouvrier R, Seeger J, De Jonghe P, Weis J, Kruttgen A, Rudnik-Schoneborn S, Bergmann C, Suter U, Zerres K, Timmerman V, Relvas JB, Senderek J (2007) Peripheral nerve demyelination caused by a mutant Rho GTPase guanine nucleotide exchange factor, frabin/FGD4. Am J Hum Genet 81:158–164

Stendel C, Roos A, Kleine H, Arnaud E, Ozcelik M, Sidiropoulos PN, Zenker J, Schupfer F, Lehmann U, Sobota RM, Litchfield DW, Luscher B, Chrast R, Suter U, Senderek J (2010) SH3TC2, a protein mutant in Charcot-Marie-Tooth neuropathy, links peripheral nerve myelination to endosomal recycling. Brain 133:2462–2474

Suter U, Snipes GJ (1995) Biology and genetics of hereditary motor and sensory neuropathies. Annu Rev Neurosci 18:45–75

Suter U, Moskow JJ, Welcher AA, Snipes GJ, Kosaras B, Sidman RL, Buchberg AM, Shooter EM (1992a) A leucine-to-proline mutation in the putative first transmembrane domain of the 22-kDa peripheral myelin protein in the trembler-J mouse. Proc Natl Acad Sci USA 89:4382–4386

Suter U, Welcher AA, Ozcelik T, Snipes GJ, Kosaras B, Francke U, Billings-Gagliardi S, Sidman RL, Shooter EM (1992b) Trembler mouse carries a point mutation in a myelin gene. Nature (Lond) 356:241–244

Takashima H, Boerkoel CF, De Jonghe P, Ceuterick C, Martin JJ, Voit T, Schroder JM, Williams A, Brophy PJ, Timmerman V, Lupski JR (2002) Periaxin mutations cause a broad spectrum of demyelinating neuropathies. Ann Neurol 51:709–715

Tazir M, Bellatache M, Nouioua S, Vallat JM (2013) Autosomal recessive Charcot-Marie-Tooth disease: from genes to phenotypes. J Peripher Nerv Syst 18:113–129

Tobler AR, Notterpek L, Naef R, Taylor V, Suter U, Shooter EM (1999) Transport of Trembler-J mutant peripheral myelin protein 22 is blocked in the intermediate compartment and affects the transport of the wild-type protein by direct interaction. J Neurosci 19:2027–2036

Vance JM, Nicholson GA, Yamaoka LH, Stajich J, Stewart CS, Speer MC, Hung WY, Roses AD, Barker D, Pericak-Vance MA (1989) Linkage of Charcot-Marie-Tooth neuropathy type 1a to chromosome 17. Exp Neurol 104:186–189

Zephir H, Stojkovic T, Latour P, Hurtevent JF, Blankaert F, Vermersch P (2005) A family with a novel frameshift mutation in the PMP22 gene (c.433_434insC) causing a phenotype of hereditary neuropathy with liability to pressure palsies. Neuromuscul Disord 15:493–497

Zuchner S, Noureddine M, Kennerson M, Verhoeven K, Claeys K, De Jonghe P, Merory J, Oliveira SA, Speer MC, Stenger JE, Walizada G, Zhu D, Pericak-Vance MA, Nicholson G, Timmerman V, Vance JM (2005) Mutations in the pleckstrin homology domain of dynamin 2 cause dominant intermediate Charcot-Marie-Tooth disease. Nat Genet 37:289–294

Chapter 7
Expression of the Transthyretin Gene in Schwann Cells and Familial Amyloidotic Polyneuropathy-Mediated Neurodegeneration

Tatsufumi Murakami and Yoshihide Sunada

Abstract Familial amyloidotic polyneuropathy (FAP) is a hereditary systemic amyloidosis, characterized by peripheral neuropathy. Amyloid derived from most types of FAP consists of transthyretin (TTR) variants with single amino-acid substitutions. Patients with FAP typically show early features of sensory polyneuropathy and autonomic neuropathy. Studies have shown that TTR is mainly produced in the liver and choroid plexus, but not in the peripheral nervous system. However, using laser capture microdissection and reverse transcription-polymerase chain reaction, our group recently discovered the expression of the TTR gene in peripheral glial cells of the dorsal root ganglia. However, the source of TTR amyloid deposits may not be restricted to this cell type, as subsequent studies have revealed that the TTR gene is also expressed in Schwann cells of peripheral nerves. Pathological studies on FAP have shown that the neuropathy is primarily axonopathy, despite numerous studies revealing demyelination and Schwann cell abnormalities. TTR amyloid deposits were observed to have a close association with Schwann cells. TTR synthesis in Schwann cells may therefore explain why TTR variants accumulate in the peripheral nervous system in FAP. In this review, we discuss a relationship between expression of the TTR gene in Schwann cells and the pathogenesis of FAP.

Keywords Amyloid • Dorsal root ganglia • Familial amyloidotic polyneuropathy • Satellite cell • Schwann cell • Transthyretin

T. Murakami (✉) • Y. Sunada
Department of Neurology, Kawasaki Medical School,
577 Matsushima, Kurashiki, Okayama 701-0192, Japan
e-mail: tatsum@med.kawasaki-m.ac.jp

K. Sango and J. Yamauchi (eds.), *Schwann Cell Development and Pathology*,
DOI 10.1007/978-4-431-54764-8_7, © Springer Japan 2014

7.1 Introduction

Familial amyloidotic polyneuropathy (FAP) is an autosomal dominantly inherited systemic amyloidosis, characterized by extracellular amyloid deposition and prominent peripheral nerve impairment (Planté-Bordeneuve and Said 2011). Andrade (1952) was the first to demonstrate early impairment of temperature and pain sensation in the feet and autonomic dysfunction in Portuguese FAP patients. FAP type 1, the most common form of FAP found in the Portuguese population, has also been detected in Japanese and Swedish populations (Andersson 1970; Araki et al. 1968). The age of onset for FAP is typically between 20 and 35 years of age, and its symptoms thereafter progress to death within 10 to 15 years. Electrophysiological diagnosis of FAP presents axonal neuropathy (Luis 1978). Sural nerve biopsy shows early loss of unmyelinated fibers and small myelinated fibers (Dyck and Lambert 1969). In FAP type 1, amyloid fibrils have been shown to be immunologically related to transthyretin (TTR) (Costa et al. 1978), in which valine is substituted by a methionine at position 30 (Tawara et al. 1983). In most cases of FAP, amyloid fibrils consist of a variant TTR with single amino-acid substitutions. According to Connors et al. (2003), more than 100 TTR amyloidogenic variants have been reported in FAP patients worldwide.

TTR is a 55,000-Da tetramer consisting of identical subunits of 127 amino-acid residues transporting retinol and thyroxine in the blood. Upon cloning of TTR cDNA in the mid-1980s (Mita et al. 1984), expression studies of this gene then followed in humans, mice, and rats via dot blot, Northern blot, and in situ hybridization (Fung et al. 1988; Mita et al. 1986; Murakami et al. 1987; Soprano et al. 1985; Wakasugi et al. 1985). The TTR gene is mainly expressed in hepatocytes of the liver and the choroid plexus of the brain (Murakami et al. 1995). TTR mRNA has also been detected in the pigmented epithelial cells of the retina (Martone et al. 1988) and α-islet cells of the pancreas (Jacobsson 1989), whereas low levels have been noted in the meninges (Blay et al. 1993). Expression of the TTR gene was not reported in the peripheral nervous system (PNS).

In FAP, TTR-containing amyloid deposits have been shown to occur throughout the PNS; however, its origin and tissue selectivity remain unknown. We have hypothesized that the TTR gene may be expressed in the PNS and have thus demonstrated that the TTR gene is expressed in peripheral glial cells of dorsal root ganglia (DRG) (Murakami et al. 2008). Subsequent findings also revealed the expression of the TTR gene in the Schwann cells of the peripheral nerves (Murakami et al. 2010). In this chapter, we speculate about the role of TTR in the PNS and its relationship to nerve impairment in FAP by reviewing our previous studies.

7.2 The TTR Gene Is Expressed in the DRG

7.2.1 Reverse Transcription-Polymerase Chain Reaction Analysis of TTR in Human, Mouse, and Rat DRG

In humans, TTR mRNA was found not to be expressed in the sural nerve, lumbar plexus, or sympathetic ganglia (Murakami et al. 2008). Surprisingly, however, it was clearly observed in DRG samples (Murakami et al. 2008). Further tests in rodent DRG confirm these results, as mouse and rat TTR mRNA was detected in the DRG, but not in the sciatic nerve.

Analyses of the anterior and posterior roots to which the DRG was attached revealed that mouse TTR mRNA was not expressed in these locations, but only in the DRG itself (Murakami et al. 2009). Furthermore, it was shown to be expressed in the L2, L3, L4, and L5 regions of the DRG (Fig. 7.1a).

Overall, the DRG may specifically express the TTR gene. Further, experiments explored the expression site of the TTR gene in DRG.

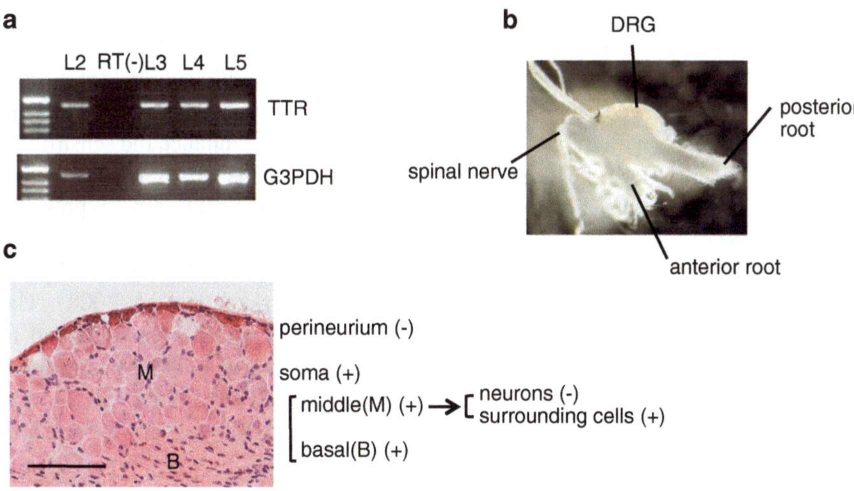

Fig. 7.1 Expression of the transthyretin (*TTR*) gene in a mouse dorsal root ganglion (DRG). **a** RT-PCR analysis of mouse DRG. RT (–) indicates no reverse transcriptase addition to L2 DRG RNA sample. **b** Mouse *DRG* sample observed under a stereomicroscope. **c** Distribution analysis of the expression of the TTR gene in a mouse DRG section stained with hematoxylin and eosin. (+) indicates TTR gene expression in the parts of DRG or cells; (–) indicates no expression of the TTR gene. *Bar* 100 μm. (From Murakami et al. 2009)

7.2.2 *In Situ Hybridization Analysis of the TTR Gene in Mouse DRG*

In situ hybridization for the TTR gene was performed in the developing mouse brain by Murakami et al. (1987). Our findings revealed intense hybridization signals in the mouse choroid plexus but none in the sensory neurons, glial cells, and perineurial cells of the DRG (Murakami et al. 2008). Additional data in the mouse showed no expression of TTR mRNA in the meninges (Murakami et al. 2009). Overall, the expression level of the TTR gene was found to be low in the mouse DRG; however, the sensitivities of in situ hybridization may have been insufficient to detect TTR mRNA in this region.

7.2.3 *Laser Capture Microdissection Followed by RT-PCR to Increase Sensitivity for the Detection of TTR in Mouse DRG*

Laser capture microdissection (LCM) is a powerful method to analyze mRNA and protein content at the cellular or tissue level (Sousa et al. 2007). Mouse lumbar DRG were dissected and frozen in liquid nitrogen and DRG sections were mounted on slides. Using LCM, specific parts of this region were collected, and total RNA was isolated followed by reverse transcription-polymerase chain reaction (RT-PCR) analysis (with the number of cycles increased to enhance the sensitivity) (Murakami et al. 2008). DRG were attached to the stumps of anterior, posterior roots, and spinal nerves (Fig. 7.1b). We first confirmed the TTR gene expression in whole DRG.

Then, TTR mRNA was found to be located in soma, middle, and basal parts of DRG, but not in the perineurium (Fig. 7.1c) (Murakami et al. 2008). The basal part included Schwann cells. Furthermore, TTR mRNA was detected in the surrounding cells of the middle part, satellite cells, and Schwann cells. However, it was not found in sensory neurons (Fig. 7.1c). No TTR mRNA was detected in sensory neurons of primary cultures of rat DRG neurons (Murakami et al. 2008).

Overall, these results clearly indicate that the TTR gene is expressed in peripheral glia, Schwann cells, and satellite cells of DRG. However, in a previous study by Sousa and Saraiva (2008), who utilized the same preparation and LCM method, expression of TTR mRNA was not detected. This result was possibly caused by the low expression of this gene in DRG; hence, our approach was to increase the number of PCR cycles to enhance the sensitivity of detection.

7.2.4 *Differences in Expression of the TTR Gene in the DRG of Humans and Rodents*

TTR mRNA was detected as a distinct band in human DRG samples but was faint in the mouse and rat (Murakami et al. 2008). Quantification of TTR mRNA confirmed these observations revealing a 19- or 33-fold increase of its expression in human compared with that of mouse or rat, respectively (Murakami et al. 2008). The differences in the expression levels of the TTR gene between the species may be caused by the differences in the promoter sequences. The human promoter of TTR contains five putative binding sites for SRY/SOX transcriptional factors within 1-kb regions. The mouse promoter does not contain such sites, and the rat promoter contains only one site. Electrophoretic mobility shift and circular permutation assays have shown that the putative SRY/SOX binding sites for the human TTR gene promoter bind the recombinant human DNA-binding box of several proteins from the SRY/SOX family (SRY, SOX9, SOX10) (Prokunina et al. 2002). SOX10 may be the candidate transcriptional factor because of its expression in tissues usually affected in FAP (Britsch et al. 2001). Thus, further studies are necessary to understand the regulation of the TTR gene in peripheral glial cells of human and rodent.

7.3 Expression of the TTR Gene in Schwann Cells

7.3.1 *Studies in the Peripheral Nerve*

Although the DRG contains Schwann cells, other areas of the PNS, such as the sciatic nerves, contain this type of glial cell (Jessen and Mirsky 2005). TTR mRNA was found to be expressed in the sciatic nerve of mice, which was revealed to be one-fifth of that of the DRG (Murakami et al. 2010). TTR protein in the endoneurium has previously been shown in the sciatic nerve of mice and is present in the extracellular matrix surrounding nerve fibers (Fleming et al. 2009). We thus postulated that TTR may be secreted from Schwann cells in the nerve.

Further, our experiments revealed that TTR mRNA was also abundant in sciatic nerves that were ligated in mice compared with the nonligated side (Fig. 7.2a). Moreover, G3PDH mRNA was also highly expressed in the ligated side, thus possibly reflecting the dedifferentiation and proliferation of Schwann cells in the ligated nerves (Dyck et al. 2005). Schwann cells often form bands of Büngner after a nerve crush (Dyck et al. 2005). Therefore, TTR may facilitate the creation of an outgrowth permissive environment.

In human peripheral nerves, we found that TTR mRNA was weakly expressed in the intercostal nerve (Fig. 7.2b). Taken together, the expression of TTR mRNA in peripheral nerves of mice and human indicates that expression of this gene may increase under pathological conditions.

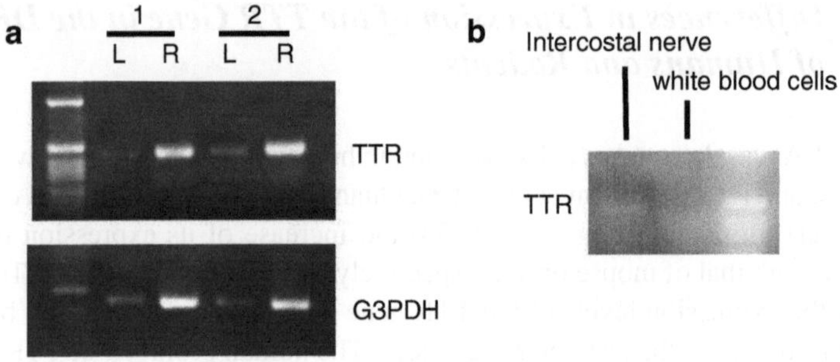

Fig. 7.2 Expression of the TTR gene in ligated mouse sciatic nerves and human intercostal nerve. **a** RT-PCR analysis of mouse TTR gene expression in ligated right sciatic nerves (*R*) and nonligated left sciatic nerves (*L*). Sciatic nerves in the unilateral leg of mice were ligated by suture and examined after 1 week. **b** RT-PCR (30 cycles) analysis of human TTR gene in the intercostal nerve. Human white blood cells were used as a negative control

7.3.2 Further Confirmation of the Expression of the TTR Gene in Cultured Schwann Cells

TTR mRNA was detected in primary cultures of Schwann cells derived from sciatic nerves of 8-day-old mice and in the immortalized murine Schwann cell line (derived from the DRG), IMS32 (Watabe et al. 1995; Murakami et al. 2010). Furthermore, we also found expression of the TTR gene in the rat Schwannoma cell line, RT4 (Freeman and Sueoka 1987; Matsumura et al. 1997) (Fig. 7.3a). Addition of forskolin (200 μM) (Sobue et al. 1986) to RT4 cell cultures was shown to induce a change in the morphology of RT4 cells to a spindle shape (Fig. 7.3b), thus demonstrating an induction of cellular differentiation. Furthermore, the level of expression of the TTR gene was decreased 96 h after the addition of forskolin (Fig. 7.3c) and was significantly lower (0.34-fold) than nonstimulated RT4 cells (Fig. 7.3d). Therefore, these data suggest that the expression of the TTR gene may be decreased by Schwann cell differentiation.

7.3.3 Transgenic Mice Carrying the Human TTR Met30 Gene in a Mouse Ttr-Null Background

Human TTR mRNA was detected in each lumbar region of the DRG and the sciatic nerve in mice (Murakami et al. 2010) carrying the human TTR Met30 gene with its cognate 6-kb upstream region in a mouse Ttr-null background (Kohno et al. 1997). Furthermore, human TTR was found in cell lysates and medium of primary cultures of Schwann cells derived from the mice, suggesting that human TTR protein is synthesized in these cells and is secreted into the medium.

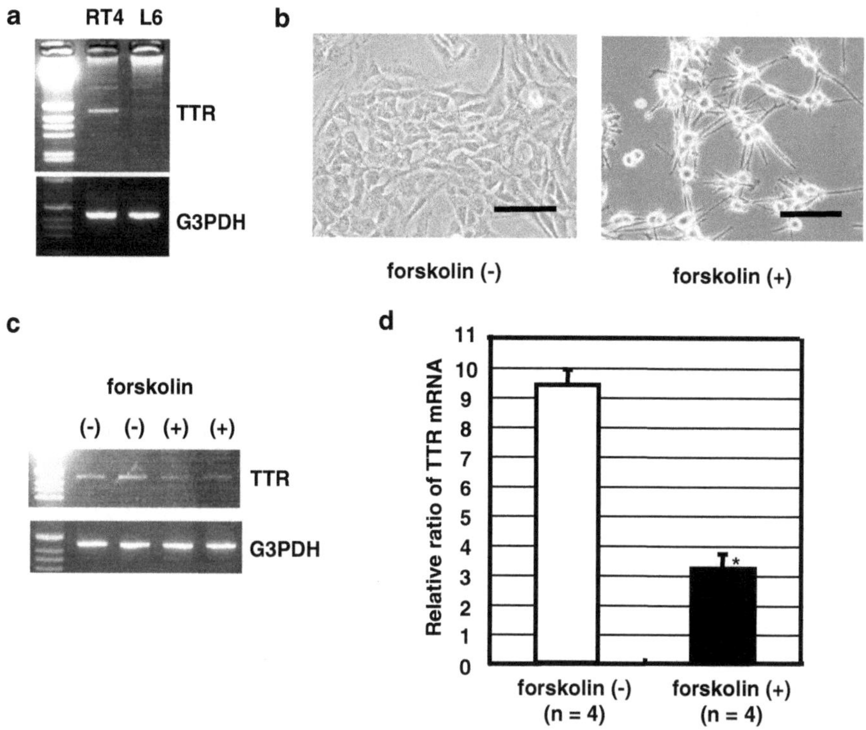

Fig. 7.3 Expression of TTR gene in RT4 cells. **a** RT-PCR (30 cycles) analysis of TTR. **b** Morphology of RT4 cells before (*left*) or after (*right*) addition of 200 μM forskolin. *Bar* 100 μm. **c** Level of expression of TTR gene 96 h after addition of forskolin. **d** Quantitative PCR (qPCR) analysis of expression of TTR gene. Values represent the mean ± SEM. $^{*}p<0.001$, Student's *t* test

7.3.4 Physiological Role of TTR Synthesis in Schwann Cells and Satellite Cells

Despite earlier reports that TTR knockout mice are phenotypically normal (Episkopou et al. 1993), later studies by Fleming et al. (2007) demonstrated impaired sensorimotor performance in this model. No pathological or electrophysiological abnormality was recognized in the peripheral nerve (Fleming et al. 2007). Furthermore, when nerve crush was performed in TTR knockout mice, the regeneration of nerve fibers, particularly unmyelinated fibers, was delayed compared with wild-type mice. In addition, in vitro experiments in the study demonstrated that TTR could directly increase neurite outgrowth without thyroxine or retinol. Retrograde transport was also impaired in TTR knockout mice (Fleming et al. 2009). In primary cultures of DRG neurons, TTR was shown to be internalized via clathrin-dependent endocytosis, with megalin as the endocytic TTR receptor, necessary for TTR neuritogenic activity (Fleming et al. 2009). Altogether, TTR synthesis in Schwann cells may be important for the maintenance of the microenvironment of endoneurium for regenerating fibers.

7.4 Pathogenesis of FAP

7.4.1 Neuropathology in FAP

Neuropathology of FAP has been well studied using sural nerve biopsies. These studies have shown axonal loss of unmyelinated and small myelinated fibers as the characteristic feature of FAP (Dyck and Lambert 1969; Guimarães et al. 1990; Thomas and King 1974). Early loss of unmyelinated axons was found on electron microscopy to correspond with the symptoms of onset of FAP, pain, and temperature impairments. TTR amyloid deposits occur in the endoneurium, near the capillary, and close to Schwann cells (Fig. 7.4) and collagen fibrils (Coimbra and Andrade 1971a; Dyck and Lambert 1969; Thomas and King 1974). However, in some nerve biopsies, TTR amyloid was not present (Coimbra and Andrade 1971b; Guimaraes et al. 1990; Said and Plante-Bordeneuve 2009). Coimbra and Andrade (1971a) have speculated that these amyloid deposits do not contribute to neuropathy because nerve fiber changes precede amyloid deposition. Destruction of endoneurial vessels is a late event in the course of FAP (Said and Planté-Bordeneuve 2009). A teased-fiber study revealed that single fibers near amyloid deposits consisted of different abnormalities, such as the distortion of myelin sheaths, segmental demyelination, and Wallerian degeneration (Said et al. 1984). The study speculated that the accumulation of amyloid-induced lesions on a single fiber may induce distal degeneration of axons, which occurs in the dying-back process. However, this mechanical theory does not explain why unmyelinated fibers are predominantly affected.

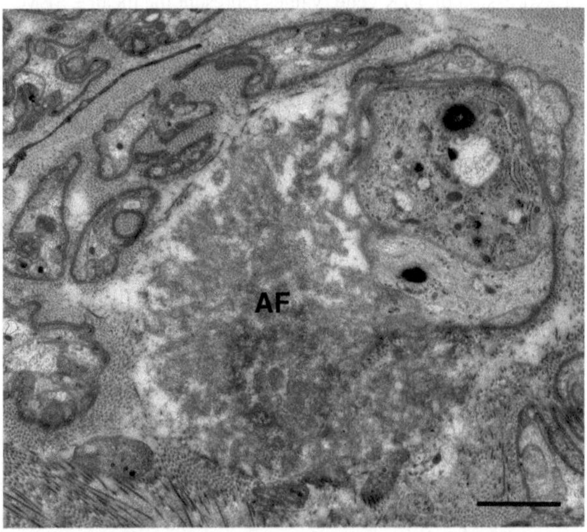

Fig. 7.4 Sural nerve in a patient with familial amyloidotic polyneuropathy (FAP) type 1. Electron micrograph of a transverse section of the sural nerve, showing deposits of amyloid fibrils (*AF*) of TTR in endoneurium associated with Schwann cell and collagen fibrils. *Bar* 2 μm

Although axonal degeneration is the primary change in FAP, Schwann cell abnormalities in this disease have also been described. Indeed, Schwann cells and myelin sheaths have been reported to be more affected than axons in myelinated fibers (Coimbra and Andrade 1971b). Schwann cell crescents contain glycogen particles, multimembranous bodies, and vacuoles. Amyloid deposits were shown to be closely associated with Schwann cells, demonstrated by their physical links with Schwann cell basal lamina (Jedrzejowska 1977; Said et al. 1984; Thomas and King 1974). A sural nerve biopsy from an asymptomatic child of a FAP type 1 parent revealed a small amount of fibrous materials about 10 nm wide with double contours in Schwann cells of myelinated fibers (Takahashi et al. 1974). In four asymptomatic children of FAP patients, frequent glycogen deposits and clusters of multimembranous bodies were noted in Schwann cell crescents of large myelinated fibers (Carvalho et al. 1976). The myelinated fibers showed infoldings of myelin sheath, irregular myelin lamination, and numerous Schmidt–Lanterman incisures.

Sural nerve biopsies represent a restricted portion of the PNS; however, the information it provides for the DRG and proximal nerve trunk may be important to understand the pathogenesis of FAP. The selective early loss of unmyelinated fibers may be attributed to a more proximal lesion such as those of DRG neurons (Dyck and Lambert 1969; Jedrzejowska 1977; Thomas and King 1974). Autopsy cases of amyloid neuropathy have provided pathological findings at these lesion sites (Hofer and Anderson 1975; Ikeda et al. 1987; Takahashi et al. 1991; Yamada et al. 1984).

Peripheral nerve abnormalities analyzed in three autopsy cases of FAP type 1 revealed nerve lesions in sciatic nerves and brachial plexuses, which exhibited a multifocal distribution of prominent interstitial edema adjacent to amyloid deposits with nerve fiber loss (Hanyu et al. 1989). In addition, the teased-fiber technique revealed segmental demyelination. Moreover, in the sural nerve, diffuse fiber loss with axonal degeneration was observed, and multifocal lesions in the proximal portions summated distally to produce polyneuropathy in FAP.

Neuropathological changes of FAP were described in two autopsy cases of FAP type 1 (Sobue et al. 1990). In the DRG, prominent loss of small sensory neurons was observed. Furthermore, amyloid deposits were noted in the endoneurium of the peripheral nerves, but more markedly in the DRG, sympathetic ganglia (SyG), and more proximal portions of nerves. In addition, Schwann and satellite cells were surrounded by amyloid fibrils and frequently showed disappearance of basement membrane. The teased-fiber technique also revealed segmental demyelination and remyelination in the proximal portion of nerve and axonal degeneration of the distal portion.

Koike et al. (2004) compared the pathological features between early- and late-onset TTR Met30 FAP in Japanese patients. Results revealed that amyloid deposition occurred throughout the DRG and SyG in both times of onset; however, amounts and neuronal loss were greater in early-onset cases.

Quantitative analyses of two autopsy cases of FAP with the Gly42 TTR mutation revealed amyloid deposition in the DRG, SyG, and in the peripheral nerves, with some accentuation in the more proximal portion (Toyooka et al. 1995). Further findings from this study showed a severe loss of predominantly small DRG neurons with the depletion of their afferent fiber in the spinal dorsal horn.

In addition, the teased-fiber technique demonstrated demyelinating fibers, as well as axonal degeneration. Moreover, under ultrastructure examination, amyloid fibrils were shown to be in direct contact with the axoplasmic membrane of demyelinating axons. Interestingly, the results of this study suggested that the presence of demyelination was not necessary for axonal destruction and that amyloid deposits along the nerve fibers (which are frequently associated with demyelination and remyelination) may not always be necessary for nerve degeneration. Overall, the study concluded that primary axonal degeneration and ganglionopathy caused by amyloid deposits may be the pathogenic mechanisms for peripheral neuropathy in FAP.

Studies in sural nerve and autopsy specimens of FAP with the Lys54 TTR mutation revealed that amyloid did not invade nerve fibers or Schwann cells despite some of their destruction (Nagasaka et al. 2009). Moreover, degeneration of the gracile fasciculus was found in the spinal cord without amyloid deposition. In FAP type 1 cases, minimal to moderate loss of myelinated fibers was found in the posterior column (Koike et al. 2004). These results suggest that the dying-back degeneration caused by ganglionopathy may occur in the PNS of FAP individuals.

7.4.2 Aggregation of TTR and Neurotoxicity

Sural nerve biopsies have revealed that extracellular spaces in close proximity to myelinated and unmyelinated nerves are filled with amyloid fibrils and an amorphous material (Coimbra and Andrade 1971a; Inoue et al. 1998). Examination of TTR and amyloid in the sural nerve of FAP patients at different stages of the disease showed that only TTR is deposited in the nerves of asymptomatic carriers (Sousa et al. 2001a). Moreover, nonfibrillar aggregates of TTR were found near Schwann cells. A subsequent study showed neuronal stress in the sural nerves at the very early stage of FAP (Sousa et al. 2001b). The expression of receptor for advanced glycation end products, proinflammatory cytokines (tumor necrosis factor-α and interleukin-1β), and the inducible form of nitric oxide synthase was increased in axons but not in Schwann cells.

Amyloidogenic TTR variants were shown to be toxic to the human neuroblastoma cell line, IMR32 (Andersson et al. 2002). Moreover, electron microscopy revealed small globular structures and protofibrils of these variants. Mature amyloid derived from FAP patients did not show a toxic effect. Using the same cell line, Reixach et al. (2004) showed that TTR monomer and nonnative oligomers (<100 kDa) were the major toxic species, although TTR amyloid fibrils and soluble TTR aggregates (>100 kDa) were not toxic.

The amyloidogenic Pro50 TTR variant was shown to cause significant calcium influx in growth cones of small-diameter TrkA-positive DRG sensory neurons (Gasperini et al. 2011). Their data suggested that activation of transient receptor potential M8 (TRPM8) channels, by the TTR variant, caused the activation of Nav1.8 voltage-gated sodium channels, leading to calcium influx through voltage-gated calcium channels. Therefore, calcium deregulation in sensory DRG neurons may be related to the pathogenesis of FAP.

Amyloids or nonfibrillar TTR deposits have not been reported in the peripheral nerves of transgenic mice expressing Met30 TTR or Leu55 TTR (Sousa et al. 2002; Kohno et al. 1997). However, crossing of Met30 TTR mice with mice lacking the main heat shock transcription factor (Hsf1), only nonfibrillar TTR deposits were detected in the sciatic nerve and DRG (Teixeira et al. 2006; Santos et al. 2010). The percentage of TTR deposition in the DRG was higher than in sciatic nerves (Santos et al. 2010), suggesting that the DRG may affect PNS tissue first in FAP. Interestingly, TTR was deposited in close proximity to satellite cells in the DRG and to Schwann cells in sciatic nerves (Santos et al. 2010). The study suggested that HSF1-regulated genes are involved in TTR tissue deposition.

Formation of intracellular aggregates of 20-nm spherules and amyloid filaments were reported in adipose tissue and brain glia in a *Drosophila* model expressing Asn14/Glu16 TTR (Pokrzywa et al. 2010). The study concluded that uptake of TTR from the circulation and its subsequent segregation into cytoplasmic arrays of nanospheres was part of a mechanism to neutralize the toxic effect of TTR. Thus, it was speculated that early formation of intracellular TTR amyloid may have been overlooked in the pathology of TTR amyloidosis because of late clinical diagnosis.

7.4.3 Liver Transplantation and Neuropathy

Because the liver is a major source of TTR variants, liver transplantation (LT) has thus been performed as a therapeutic strategy in the early course of FAP (Benson 2013; Suhr 2003). LT halts the progress of symptoms in most patients (Planté-Bordeneuve and Said 2011; Yamamoto et al. 2007). The latter study revealed that sensory disturbances improved in 40 % of patients, remained unchanged in 49 %, and deterioration of the condition was experienced in 10.7 % of patients. Furthermore, motor disturbances were improved in 12 % of patients, remained unchanged in 83 %, and deteriorated in 5 %. In some patients, neuropathic progression was reported after LT; this effect may be a result of the continuous deposition of TTR secreted from Schwann cells or the choroid plexus. Aging may also be an important factor determining the progression of neuropathy after LT (Koike et al. 2012).

7.4.4 Schwann Cell Hypothesis

Several hypotheses may explain neurodegeneration in FAP (Said 2003; Sousa and Saraiva 2003): (1) the mechanical effect, involving multiple compressions of the nervous tissues by amyloid; (2) nerve ischemia; (3) toxic effect of nonfibrillar TTR aggregates; and (4) lesions in DRG neurons or Schwann cells. Because the TTR gene is expressed in Schwann cells (Murakami et al. 2010), the Schwann cell hypothesis may be regarded as an initial trigger for neurodegeneration. However, at the late stage of FAP, a cumulative effect of all the neurodegenerative mechanisms occurring in FAP may be responsible for nerve fiber degeneration.

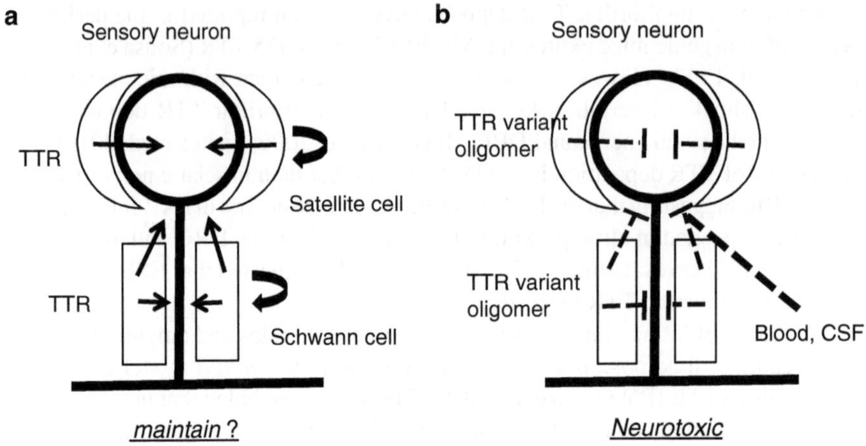

Fig. 7.5 Hypothesis of the involvement of the DRG in FAP. Schema demonstrate (**a**) satellite cells and Schwann cells secreting TTR and other substances, maintaining the microenvironment of the DRG and (**b**) TTR variants (at the onset of FAP) from peripheral glia, blood, and cerebral spinal fluid (*CSF*) possibly forming TTR oligomers, which exert toxic effects on sensory neurons. (Modified from Murakami et al. 2009)

The DRG may be the first affected target for FAP because the initial symptom is usually sensory (Planté-Bordeneuve and Said 2011), amyloid is markedly deposited in the DRG (as evidenced in postmortem studies) (Sobue et al. 1990; Toyooka et al. 1995), and the TTR gene is expressed in the peripheral glial cells of the DRG (Murakami et al. 2008). In addition, the DRG does not contain a blood–brain barrier and thus is constantly exposed to the circulating blood (Hanani 2005). TTR in cerebrospinal fluid (CSF) may also be transported to the DRG through the roots (Said 2003). The satellite cell (Schwann cell homologue) is closely attached to DRG sensory neurons and maintains the microenvironment (Hanani 2005). Satellite cells provide neurotrophic factors, nutrients, and other substances, including TTR, thus protecting sensory neurons (Hanani 2005; Murakami et al. 2008) (Fig. 7.5a). Satellite cell dysfunction may relate to sensory neuron involvement in FAP. TTR aggregates from blood, CSF, and peripheral glia (Schwann cells and satellite cells) may be toxic to sensory neurons (Fleming et al. 2007, 2009) (Fig. 7.5b). Sensory axonal degeneration subsequently occurs as a dying-back neuropathy (Nagasaka et al. 2009; Toyooka et al. 1995) (Fig. 7.6). However, the reason why small sensory neurons are more affected remains unknown.

Subsequent motor symptoms may occur in the distal portions of the lower extremities. No primary abnormalities in motor neurons of the spinal cord were observed in FAP autopsies (Sobue et al. 1990; Toyooka et al. 1995); therefore, axonal degeneration by multiple compressions, or the toxic effect of TTR aggregates in the peripheral nerves, particularly the proximal nerve trunk, must be considered as a pathogenesis (Fig. 7.6). Dedifferentiated and proliferated Schwann cells, caused by sensory axonal degeneration, may increase the synthesis of TTR in the endoneurium, leading to TTR aggregates and amyloid deposits and resulting in the degeneration of motor axons (Toyooka et al. 1995).

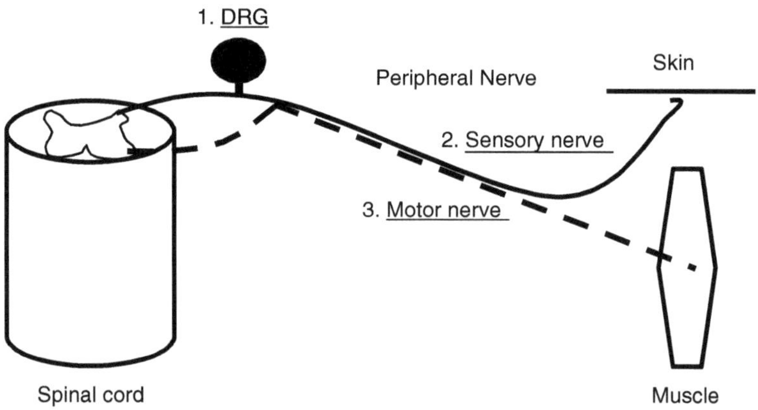

Fig. 7.6 Hypothesis of a pathogenic mechanism of neuropathy in FAP. Schema demonstrate the DRG to be initially involved with TTR aggregates or amyloid (*1*), followed by sensory axonal neuropathy occurring in a dying-back degeneration (*2*), then motor axon impairment by TTR aggregates or amyloid (*3*)

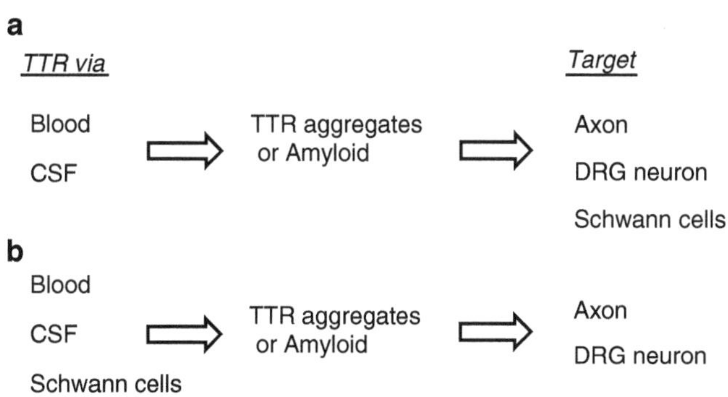

Fig. 7.7 Theories of the targets of TTR aggregates or amyloid. **a** In the previous theory, Schwann cells are not considered as important for the pathogenesis of FAP. **b** Schwann cell hypothesis whereby Schwann cells may play an important role for neurodegeneration in FAP

7.5 Conclusion

Our studies revealed that TTR mRNA was found in human DRG. The site of expression of this gene in the DRG was difficult to determine because of its low expression in the rodent PNS. However, laser capture microdissection together with RT-PCR revealed that the TTR gene in the DRG was expressed in peripheral glia. Subsequent studies confirmed the expression of this gene in Schwann cells. In FAP, neuropathy is primarily exhibited via axonal degeneration, and Schwann cells have been considered to be bystanders (Fig. 7.7a). However, previous studies have described demyelination of the peripheral nerves and abnormalities of Schwann cells.

Schwann cells and satellite cells are shown to be in close contact with TTR amyloid deposits. After the discovery of TTR synthesis in Schwann cells, the Schwann cell hypothesis was thus considered as a pathogenesis of FAP (Fig. 7.7b). This hypothesis may explain why TTR variants are preferentially deposited in the PNS. Further studies on TTR metabolism in Schwann cells, and a suitable transgenic mouse model for neuropathy of FAP, are necessary to reveal the mechanism of neurodegeneration of FAP.

References

Andersson R (1970) Hereditary amyloidosis with polyneuropathy. Acta Med Scand 188:85–94

Andersson K, Olofsson A, Nielsen EH, Svehag SE, Lundgren E (2002) Only amyloidogenic intermediates of transthyretin induce apoptosis. Biochem Biophys Res Commun 294:309–314

Andrade C (1952) A peculiar form of peripheral neuropathy; familiar atypical generalized amyloidosis with special involvement of the peripheral nerves. Brain 75:408–427

Araki S, Mawatari S, Ohta M, Nakajima A, Kuroiwa Y (1968) Polyneuritic amyloidosis in a Japanese family. Arch Neurol 18:593–602

Benson MD (2013) Liver transplantation and transthyretin amyloidosis. Muscle Nerve 47:157–162

Blay P, Nilsson C, Owman C, Aldred A, Schreiber G (1993) Transthyretin expression in the rat brain: effect of thyroid functional state and role in thyroxine transport. Brain Res 632:114–120

Britsch S, Goerich DE, Riethmacher D, Peirano RI, Rossner M, Nave K-A, Birchmeier C, Wegner M (2001) The transcription factor Sox10 is a key regulator of peripheral glial development. Genes Dev 15:66–78

Carvalho J, Coimbra A, Andrade C (1976) Peripheral nerve fibre changes in asymptomatic children of patients with familial amyloid polyneuropathy. Brain 99:1–10

Coimbra A, Andrade C (1971a) Familial amyloid polyneuropathy: an electron microscope study of the peripheral nerve in five cases. I. Interstitial changes. Brain 94:199–206

Coimbra A, Andrade C (1971b) Familial amyloid polyneuropathy: an electron microscope study of the peripheral nerve in five cases. II. Nerve fibre changes. Brain 94:207–212

Connors LH, Lim A, Prokaeva T, Roskens VA, Costello CE (2003) Tabulation of human transthyretin (TTR) variants, 2003. Amyloid 10:160–184

Costa PP, Figueira AS, Bravo FR (1978) Amyloid fibril protein related to prealbumin in familial amyloidotic polyneuropathy. Proc Natl Acad Sci USA 75:4499–4503

Dyck PJ, Lambert EH (1969) Dissociated sensation in amylidosis. Compound action potential, quantitative histologic and teased-fiber, and electron microscopic studies of sural nerve biopsies. Arch Neurol 20:490–507

Dyck PJ, Dyck PJB, Engelstad J (2005) Pathologic alterations of nerves. In: Dyck PJ, Thomas PK (eds) Peripheral neuropathy, 4th edn. Elsevier, Philadelphia, pp 733–829

Episkopou V, Maeda S, Nishiguchi S, Shimada K, Gaitanaris GA, Gottesman ME, Robertson EJ (1993) Disruption of the transthyretin gene results in mice with depressed levels of plasma retinol and thyroid hormone. Proc Natl Acad Sci USA 90:2375–2379

Fleming CE, Saraiva MJ, Sousa MM (2007) Transthyretin enhances nerve regeneration. J Neurochem 103:831–839

Fleming CE, Mar FM, Franquinho F, Saraiva MJ, Sousa MM (2009) Transthyretin internalization by sensory neurons is megalin mediated and necessary for its neuritogenic activity. J Neurosci 29:3220–3232

Freeman MR, Sueoka N (1987) Induction and segregation of glial intermediate filament expression in the RT4 family of peripheral nervous system cell lines. Proc Natl Acad Sci USA 84:5808–5812

Fung WP, Thomas T, Dickson PW, Aldred AR, Milland J, Dziadek M, Power B, Hudson P, Schreiber G (1988) Structure and expression of the rat transthyretin (prealbumin) gene. J Biol Chem 263:480–488

Gasperini RJ, Hou X, Parkington H, Coleman H, Klaver DW, Vincent AJ, Foa LC, Small DH (2011) TRPM8 and Nav1.8 sodium channels are required for transthyretin-induced calcium influx in growth cones of small-diameter TrkA-positive sensory neurons. Mol Neurodegener 6:19

Guimarães A, Viana Pinheiro A, Leite I (1990) Sural nerve biopsy in familial amyloidotic polyneuropathy: a morphological and morphometric polyneuropathy. In: Natvig JB, Forre O, Hasuby G et al (eds) Amyloid and amyloidosis. Kluwer, London, pp 493–498

Hanani M (2005) Satellite glial cells in sensory ganglia: from form to function. Brain Res Rev 48:457–476

Hanyu N, Ikeda S, Nakadai A, Yanagisawa N, Powell HC (1989) Peripheral nerve pathological findings in familial amyloid polyneuropathy: a correlative study of proximal sciatic nerve and sural nerve lesions. Ann Neurol 25:340–350

Hofer PA, Anderson R (1975) Postmortem findings in primary familial amyloidosis with polyneuropathy. Acta Pathol Microbiol Scand A 83:309–322

Ikeda S, Hanyu N, Hongo M, Yoshioka J, Oguchi H, Yanagisawa N, Kobayashi T, Tsukagoshi H, Ito N, Yokota T (1987) Hereditary generalized amyloidosis with polyneuropathy. Clinicopathological study of 65 Japanese patients. Brain 110:315–337

Inoue S, Kuroiwa M, Saraiva MJ, Guimarães A, Kisilevsky R (1998) Ultrastructure of familial amyloid polyneuropathy amyloid fibrils: examination with high-resolution electron microscopy. J Struct Biol 124:1–12

Jacobsson B (1989) In situ localization of transthyretin-mRNA in the adult human liver, choroid plexus and pancreatic islets and in endocrine tumours of the pancreas and gut. Histochemistry 91:299–304

Jedrzejowska H (1977) Some histological aspects of amyloid polyneuropathy. Acta Neuropathol 37:119–125

Jessen KR, Mirsky R (2005) The origin and development of glial cells in peripheral nerves. Nat Rev 6:671–682

Kohno K, Palha JA, Miyakawa K, Saraiva MJ, Ito S, Mabuchi T, Blaner WS, Iijima H, Tsukahara S, Episkopou V, Gottesman ME, Shimada K, Takahashi K, Yamamura K, Maeda S (1997) Analysis of amyloid deposition in a transgenic mouse model of homozygous familial amyloidotic polyneuropathy. Am J Pathol 150:1497–1508

Koike H, Misu K, Sugiura M, Iijima M, Mori K, Yamamoto M, Hattori N, Mukai E, Ando Y, Ikeda S, Sobue G (2004) Pathology of early- vs. late-onset TTR Met30 familial amyloid polyneuropathy. Neurology 63:129–138

Koike H, Hashimoto R, Tomita M, Kawagashira Y, Iijima M, Nakamura T, Watanabe H, Kamei H, Kiuchi T, Sobue G (2012) Impact of aging on the progression of neuropathy after liver transplantation in transthyretin Val30Met amyloidosis. Muscle Nerve 46:964–970

Luis ML (1978) Electroneurophysiological studies in familial amyloid polyneuropathy–Portuguese type. J Neurol Neurosurg Psychiatry 41:847–850

Martone RL, Herbert J, Dwork A, Schon EA (1988) Transthyretin is synthesized in the mammalian eye. Biochem Biophys Res Commun 151:905–912

Matsumura K, Chiba A, Yamada H, Fukuta-Ohi H, Fujita S, Endo T, Kobata A, Anderson LV, Kanazawa I, Campbell KP, Shimizu T (1997) A role of dystroglycan in schwannoma cell adhesion to laminin. J Biol Chem 272:13904–13910

Mita S, Maeda S, Shimada K, Araki S (1984) Cloning and sequence analysis of cDNA for human prealbumin. Biochem Biophys Res Commun 124:558–564

Mita S, Maeda S, Shimada K, Araki S (1986) Analyses of prealbumin mRNAs in individuals with familial amyloidotic polyneuropathy. J Biochem 100:1215–1222

Murakami T, Yasuda Y, Mita S, Maeda S, Shimada K, Fujimoto T, Araki S (1987) Prealbumin gene expression during mouse development studied by in situ hybridization. Cell Differ 22:1–9

Murakami T, Uchino M, Ando M (1995) Genetic abnormalities and pathogenesis of familial amyloidotic polyneuropathy. Pathol Int 45:1–9
Murakami T, Ohsawa Y, Sunada Y (2008) The transthyretin gene is expressed in human and rodent dorsal root ganglia. Neurosci Lett 436:335–339
Murakami T, Ohsawa Y, Sunada Y (2009) Reevaluation of transthrein gene expression in the nervous system: the expression in dorsal root ganglia. Peripher Nerv 20:30–35
Murakami T, Ohsawa Y, Zhenghua L, Yamamura K, Sunada Y (2010) The transthyretin gene is expressed in Schwann cells of peripheral nerves. Brain Res 1348:222–225
Nagasaka T, Togashi S, Watanabe H, Iida H, Nagasaka K, Nakamura Y, Miwa M, Kobayashi F, Shindo K, Shiozawa Z (2009) Clinical and histopathological features of progressive-type familial amyloidotic polyneuropathy with TTR Lys54. J Neurol Sci 276:88–94
Planté-Bordeneuve V, Said G (2011) Familial amyloid polyneuropathy. Lancet Neurol 10:1086–1097
Pokrzywa M, Dacklin I, Vestling M, Hultmark D, Lundgren E, Cantera R (2010) Uptake of aggregating transthyretin by fat body in a *Drosophila* model for TTR-associated amyloidosis. PLoS One 5:e14343
Prokunina L, Gaspari M, Thyén C (2002) Of mice and men: the case of differential regulation of the TTR gene. In: Abstracts of the 5th international symposium on familial amyloidotic polyneuropathy and other transthyretin related disorders, Matsumoto, 24–27 Sept 2002
Reixach N, Deechongkit S, Jiang X, Kelly JW, Buxbaum JN (2004) Tissue damage in the amyloidoses: transthyretin monomers and nonnative oligomers are the major cytotoxic species in tissue culture. Proc Natl Acad Sci USA 101:2817–2822
Said G (2003) Familial amyloid polyneuropathy: mechanisms leading to nerve degeneration. Amyloid 10(suppl 1):7–12
Said G, Planté-Bordeneuve V (2009) Familial amyloid polyneuropathy: a clinico-pathologic study. J Neurol Sci 284:149–154
Said G, Ropert A, Faux N (1984) Length-dependent degeneration of fibers in Portuguese amyloid polyneuropathy: a clinicopathologic study. Neurology 34:1025–1032
Santos SD, Fernandes R, Saraiva MJ (2010) The heat shock response modulates transthyretin deposition in the peripheral and autonomic nervous systems. Neurobiol Aging 31:280–289
Sobue G, Shuman S, Pleasure D (1986) Schwann cell responses to cyclic AMP: proliferation, change in shape, and appearance of surface galactocerebroside. Brain Res 362:23–32
Sobue G, Nakao N, Murakami K, Yasuda T, Sahashi K, Mitsuma T, Sasaki H, Sakaki Y, Takahashi A (1990) Type I familial amyloid polyneuropathy. A pathological study of the peripheral nervous system. Brain 113:903–919
Soprano DR, Herbert J, Soprano KJ, Schon EA, Goodman DS (1985) Demonstration of transthyretin mRNA in the brain and other extrahepatic tissues in the rat. J Biol Chem 260:11793–11798
Sousa MM, Saraiva MJ (2003) Neurodegeneration in familial amyloid polyneuropathy: from pathology to molecular signaling. Prog Neurobiol 71:385–400
Sousa MM, Saraiva MJ (2008) Transthyretin is not expressed by dorsal root ganglia cells. Exp Neurol 214:362–365
Sousa MM, Cardoso I, Fernandes R, Guimarães A, Saraiva MJ (2001a) Deposition of transthyretin in early stages of familial amyloidotic polyneuropathy: evidence for toxicity of nonfibrillar aggregates. Am J Pathol 159:1993–2000
Sousa MM, Du Yan S, Fernandes R, Guimarães A, Stern D, Saraiva MJ (2001b) Familial amyloid polyneuropathy: receptor for advanced glycation end products-dependent triggering of neuronal inflammatory and apoptotic pathways. J Neurosci 21:7576–7586
Sousa MM, Fernandes R, Palha JA, Taboada A, Vieira P, Saraiva MJ (2002) Evidence for early cytotoxic aggregates in transgenic mice for human transthyretin Leu55Pro. Am J Pathol 161:1935–1948
Sousa JC, Cardoso I, Marques F, Saraiva MJ, Palha JA (2007) Transthyretin and Alzheimer's disease: where in the brain? Neurobiol Aging 28:713–718

Suhr OB (2003) Impact of liver transplantation on familial amyloidotic polyneuropathy (FAP) patients' symptoms and complications. Amyloid 10(suppl 1):77–83

Takahashi K, Nakamura H, Ishitobi K, Harada Y (1974) Ultrastrurcture of sural nerves of amyloid neuropathy with special reference to an asymptomatic sibling. Clin Neurol 14:266–274

Takahashi K, Yi S, Kimura Y, Araki S (1991) Familial amyloidotic polyneuropathy type 1 in Kumamoto, Japan: a clinicopathologic, histochemical, immunohistochemical, and ultrastructural study. Hum Pathol 22:519–527

Tawara S, Nakazato M, Kangawa K, Matsuo H, Araki S (1983) Identification of amyloid prealbumin variant in familial amyloidotic polyneuropathy (Japanese type). Biochem Biophys Res Commun 116:880–888

Teixeira PF, Cerca F, Santos SD, Saraiva MJ (2006) Endoplasmic reticulum stress associated with extracellular aggregates. Evidence from transthyretin deposition in familial amyloid polyneuropathy. J Biol Chem 281:21998–22003

Thomas PK, King RH (1974) Peripheral nerve changes in amyloid neuropathy. Brain 97:395–406

Toyooka K, Fujimura H, Ueno S, Yoshikawa H, Kaido M, Nishimura T, Yorifuji S, Yanagihara T (1995) Familial amyloid polyneuropathy associated with transthyretin Gly42 mutation: a quantitative light and electron microscopic study of the peripheral nervous system. Acta Neuropathol 90:516–525

Wakasugi S, Maeda S, Shimada K, Nakashima H, Migita S (1985) Structural comparisons between mouse and human prealbumin. J Biochem 98:1707–1714

Watabe K, Fukuda T, Tanaka J, Honda H, Toyohara K, Sakai O (1995) Spontaneously immortalized adult mouse Schwann cells secrete autocrine and paracrine growth-promoting activities. J Neurosci Res 41:279–290

Yamada M, Hatakeyama S, Tsukagoshi H (1984) Peripheral and autonomic nerve lesions in systemic amyloidosis. Three pathological types of amyloid polyneuropathy. Acta Pathol Jpn 34:1251–1266

Yamamoto S, Wilczek HE, Nowak G, Larsson M, Oksanen A, Iwata T, Gjertsen H, Söderdahl G, Wikström L, Ando Y, Suhr OB, Ericzon BG (2007) Liver transplantation for familial amyloidotic polyneuropathy (FAP): a single-center experience over 16 years. Am J Transplant 7:2597–2604

Chapter 8
Node of Ranvier Disruption: A Key Pathophysiology in Immune-Mediated Neuropathies

Keiichiro Susuki

Abstract One important role of Schwann cells is to promote clustering of voltage-gated Na^+ channels on the axolemma at the nodes of Ranvier. Recent cumulative evidence demonstrates that disruption of nodal Na^+ channel clusters is the key pathophysiology in immune-mediated neuropathies such as Guillain–Barré syndrome. Autoimmune attack against either axons or myelin can disturb neuron–Schwann cell interactions at and near nodes, disrupting Na^+ channel clusters and resulting in nerve conduction block. Furthermore, the proteins highly accumulated at and near nodes can be a direct target of autoimmune reactions. This chapter focuses on dysfunction and disruption of nodes of Ranvier during autoimmune neuropathies.

Keywords Autoantibody • Autoimmune neuropathies • Axon–glial interactions • Complement • Na^+ channel • Node of Ranvier • Schwann cells

Abbreviations

AIDP	Acute inflammatory demyelinating polyradiculoneuropathy
AMAN	Acute motor axonal neuropathy
Caspr	Contactin-associated protein
CIDP	Chronic inflammatory demyelinating polyradiculoneuropathy
EAN	Experimental allergic neuritis
GBS	Guillain–Barré syndrome
Kv	Voltage-gated K^+

K. Susuki (✉)
Department of Neuroscience, Baylor College of Medicine,
One Baylor Plaza, Room S611-MS: BCM295, Houston, TX 77030, USA

Department of Neuroscience, Cell Biology, and Physiology,
Wright State University 3640 Colonel Glenn Hwy., Dayton, OH 45435, USA
e-mail: susuki@bcm.edu

K. Sango and J. Yamauchi (eds.), *Schwann Cell Development and Pathology*,
DOI 10.1007/978-4-431-54764-8_8, © Springer Japan 2014

MAC	Membrane attack complex
NF	Neurofascin
Nav	Voltage-gated Na^+
PNS	Peripheral nervous system

8.1 Introduction

Schwann cells are one key component of proper peripheral nervous system (PNS) function. The Schwann cells communicate with neurons to form a myelin sheath surrounding axons, to promote specific membrane domains along the axons, and to support axonal integrity (Sherman and Brophy 2005; Pereira et al. 2012). The main functions of the PNS are to transmit information both from the brain to the body and from the body to the brain. The information must be transmitted very rapidly over long distances up to 1 m with minimal energy. Vertebrates have solved this problem by creating myelin, multilamellar structures produced by Schwann cells that wrap around PNS axons. In myelinated nerve fibers, the axons are mostly ensheathed and insulated by serially arranged myelin along their entire length. In addition to forming myelin sheaths, Schwann cells actively promote the formation of distinct axonal membrane domains, such as the nodes of Ranvier, which are short gaps between two adjacent myelin segments (Poliak and Peles 2003; Salzer 2003). At the nodal axolemma, voltage-gated ion channels, cell adhesion molecules, and cytoskeletal and scaffolding proteins form molecular complexes, which together are crucial for regenerating action potentials. Because the internodal axons (the segments between nodes) are covered and insulated by myelin sheaths, the action potentials are regenerated only at the nodes, allowing rapid and efficient nerve conduction in a saltatory manner.

How is nerve conduction disturbed during PNS disease processes? Because myelinating Schwann cells are critical for axonal membrane domain organizations and action potential propagation, it is conceivable that changes in Schwann cells could cause significant damage to PNS structures and functions. In addition to the loss of insulation by myelin defects, the disruption of nodes of Ranvier is a critical factor to cause nerve conduction failure and neurological symptoms in various disease conditions involving the PNS, such as immune-mediated neuropathies. Because the formation and maintenance of nodes depend on neuron–glia interactions, both Schwann cell/myelin defects and axonal damage can disrupt nodal voltage-gated Na+ (Nav) channel clusters, causing nerve conduction failure. This chapter reviews recent evidence for nodal dysfunction or disruption during immune-mediated neuropathies. Nodes of Ranvier are involved in the pathophysiology of immune-mediated neuropathies in multiple ways: (1) nodes are selectively attacked by autoimmune reactions; (2) nodes are damaged as a consequence of demyelination; and (3) autoimmunity against nodes may further exacerbate demyelinating neuropathies.

8.2 Distinct Axonal Domains at and near Nodes of Ranvier

Myelinated axons are divided into multiple distinct membrane domains including the nodes of Ranvier, paranodes, juxtaparanodes, and internodes (Fig. 8.1a). Each domain is characterized by specific molecular complexes (Fig. 8.1b) (Poliak and Peles 2003; Salzer 2003). At the nodes, the Nav channels accumulate at the nodal axolemma to regenerate the action potentials. The submembranous scaffolding protein ankyrinG binds with Nav channels and the cell adhesion molecule neurofascin (NF) 186, forming a core molecular complex at nodal axons. During PNS

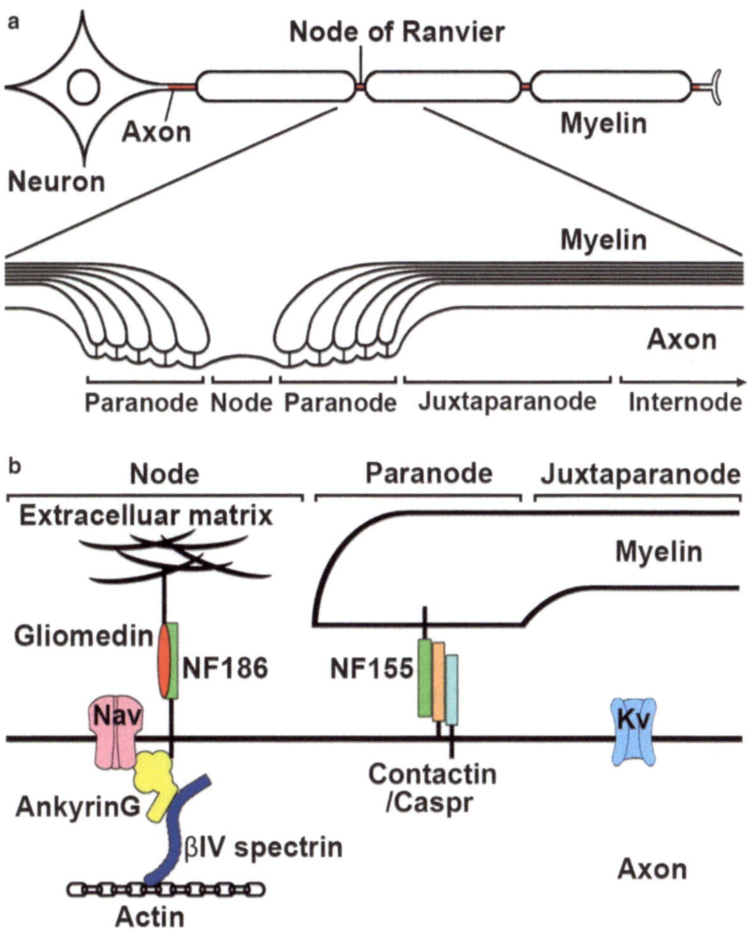

Fig. 8.1 Structures and molecular composition at and near nodes of Ranvier. **a** Cartoon illustrates the structures of the myelinated nerve fiber and axonal subdomains: nodes of Ranvier, paranode, juxtaparanodes, and internode. **b** Schematic presentation showing molecular organization at nodes, paranodes, and juxtaparanodes. (Adapted from Uncini et al. 2013, with permission)

development, the Schwann cell-derived molecules gliomedin and NrCAM are secreted and incorporated into the extracellular matrix at the edge of forming myelin segments where they bind with NF186 on the axolemma and promote its clustering (Eshed et al. 2005; Feinberg et al. 2010). This interaction between the extracellular matrix and NF186 is the leading mechanism proposed to form PNS nodes (Feinberg et al. 2010; Susuki et al. 2013). At the flanking paranodes, neurons and Schwann cells form septate-like junctions mediated by a tripartite cell adhesion molecule complex including axonal contactin and contactin-associated protein (Caspr) and glial NF155. The paranodal junctions work as a diffusion barrier and restrict the mobility of nodal molecules, providing a second mechanism to form PNS nodes (Feinberg et al. 2010). Then, two Nav channel clusters at the edge of approaching myelin merge as mature nodes between adjacent paranodal junctions. Finally, ankyrinG connects the nodal molecular complex to the actin cytoskeleton through the submembranous cytoskeletal protein βIV spectrin for further stabilization. The regions next to the paranodes, located under the myelin sheaths, are called juxtaparanodes. Paranodal junctions also restrict the mobility of proteins localized at juxtaparanodes, including voltage-gated K^+ (Kv) channels. The juxtaparanodal Kv channels are thought to mainly act as an active damper of re-entrant excitation to help in restoring and maintaining the internodal resting potential (reviewed in Poliak and Peles 2003).

8.3 Acute Motor Axonal Neuropathy

Immune-mediated neuropathies include various conditions with distinct clinical manifestations. One good example is Guillain–Barré syndrome (GBS) (for details, see Yuki and Hartung 2012). The reported incidence of GBS ranges from 0.89 to 1.89 (median, 1.11) cases per 100,000 person-years in North America and Europe (Sejvar et al. 2011). The main symptom of GBS is bilateral and relatively symmetrical limb weakness. In addition, GBS patients may develop numbness, paresthesia, or pain in the limbs, facial weakness, double vision, and respiratory failure. These symptoms reach the nadir within 4 weeks, generally following a monophasic course. The first-line treatment for GBS during the acute phase is intravenous immunoglobulin, which is thought to neutralize pathogenic autoantibodies, or plasma exchange to remove circulating autoantibodies. Despite the monophasic course of the illness and modern immunomodulative therapies, the prognosis of GBS is not favorable. Up to 20 % of patients remain severely disabled and approximately 5 % die (Hughes et al. 2007). The economic cost of GBS is substantial: the estimated annual cost is $1.7 billion in the United States, largely because of disability and death (Frenzen 2008).

GBS is divided into two subtypes, an axonal form (acute motor axonal neuropathy, AMAN) and a demyelinating form (acute inflammatory demyelinating polyradiculoneuropathy, AIDP) (Yuki and Hartung 2012). The immunopathogenesis based on molecular mimicry between infectious pathogens and peripheral nerves is well established in AMAN. Two thirds of GBS cases are preceded by symptoms of upper respiratory tract or gastrointestinal infection. The most frequently identified infectious pathogen is *Campylobacter jejuni* (30 % of GBS cases), a common bacteria that can cause enteritis. Lipo-oligosaccharides of *C. jejuni* isolated from GBS patients have structures identical to gangliosides such as GM1 or GD1a. Gangliosides are a group of glycosphingolipids composed of a ceramide attached to a sugar chain containing *N*-acetylneuraminic acid linked to an oligosaccharide core. These gangliosides are highly enriched at and near nodes and have various neurobiological functions that may include maintenance of the axon, myelin integrity, or stabilization of axon–glial interactions (Sheikh et al. 1999; Yamashita et al. 2005; Susuki et al. 2007a). Most patients with AMAN have serum IgG antibodies against gangliosides such as GM1 or GD1a (Willison and Yuki 2002). Thus, the immunopathogenesis of AMAN is considered to be as follows: (1) *C. jejuni* infection induces production of antibodies against lipooligosaccharides on the bacteria surface; (2) these antibodies cross-react to the gangliosides highly enriched in PNS; and (3) bound autoantibodies induce immune reactions causing peripheral nerve injury.

8.4 Nodal Dysfunction/Disruption in AMAN

How do anti-ganglioside antibodies induce AMAN? The autoimmune processes induced by these anti-ganglioside antibodies specifically target nodes of Ranvier. An early pathological feature in AMAN patients is widening of the nodes of Ranvier with no or little demyelination in the ventral roots (Griffin et al. 1996). The affected nodal axolemma is coated with activation products of complement, a key component of the innate immune systems (Hafer-Macko et al. 1996a). Thus, it has been speculated that the central pathophysiology in AMAN is the disruption of the nodes of Ranvier mediated by anti-ganglioside antibodies and complement; this was further confirmed by animal models. Rabbits immunized with GM1 ganglioside develop a high titer of IgG anti-GM1 antibodies and display flaccid limb weakness, identical to human AMAN patients (Yuki et al. 2001). Consistent with the pathological findings in human AMAN patients, deposition of IgG and complement products on abnormally lengthened nodes were observed in ventral roots from paralyzed rabbits during the acute phase (Fig. 8.2a, b) (Susuki et al. 2003, 2007b). Furthermore, Nav channel immunostaining disappeared as complement accumulated but without direct contact of macrophages (Susuki et al. 2007b). Other molecular components including the nodal extracellular matrix, paranodal junctions, and nodal cytoskeletal scaffolds were also disrupted in association with complement deposition, suggesting that the Nav channel clusters are destabilized by loss of these mechanisms for node formation. The initial nodal disruption may be repaired rapidly. Similar to the clinical course in human AMAN patients, the paralyzed rabbits generally start to recover in 2–4 weeks from the onset of the neurological illness. During the recovery phase, complement deposition was reduced, and nodal and paranodal molecules were again clustered on both sides of the affected nodes (Fig. 8.2a, b) (Susuki et al. 2007b). Binary Nav channel clusters then appear to fuse, allowing the reformation of nodes of Ranvier. However, if the local immune

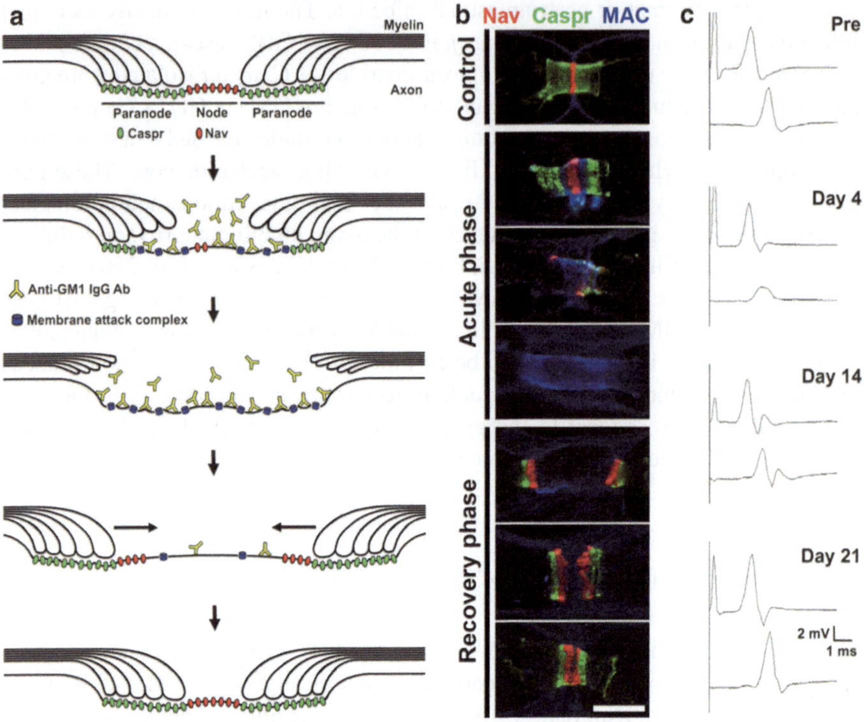

Fig. 8.2 Immune-mediated nodal disruption in acute motor axonal neuropathy (AMAN). **a** Cartoon shows the time course (from *top* to *bottom*) of nodal disruption mediated by IgG anti-GM1 antibodies and complement in AMAN. **b** Ventral root from control rabbit (*top*) or AMAN rabbit associated with IgG anti-GM1 antibodies. During the acute phase (*middle*), deposition of MAC (*blue*) occurs at the nodes first, then extends to the paranodes. Clusters of nodal Nav channels (*red*) or paranodal Caspr (*green*) are destroyed and eventually disappear. During the early recovery phase (*bottom*; 2 weeks after the onset of neurological disease), the intensity of MAC staining is reduced, and two adjacent Nav channel clusters associated with Caspr appear to fuse. *Bar* 10 μm. **c** Serial nerve conduction study in rat tibial nerve. IgG anti-ganglioside antibody was injected halfway between the ankle and knee. The nerve is stimulated at the ankle or knee, and the compound muscle action potentials are recorded from the plantar muscle. Before injection, no apparent difference is seen between the waveforms after stimulation at the ankle (*top*) and knee (*bottom*). After antibody injection (day 4), the amplitude after stimulation at the knee is abnormally reduced with no temporal dispersion, suggesting the presence of a nerve conduction block at the site of anti-ganglioside antibody injection. The amplitude after proximal stimulation returned to normal by 21 days after injection. (Figure adapted from previous publications with permission: **a**, Uncini et al. 2013; **b**, Susuki et al. 2007b; **c**, Susuki et al. 2012)

reaction progresses, axonal degeneration develops, resulting in slow recovery or even permanent neurological deficits (reviewed in Uncini et al. 2013).

The nodal disruption mediated by anti-ganglioside antibodies was further confirmed by passive transfer experiments. These passive transfer models provided more insights into the molecular mechanisms of nodal disruption as well as the physiology of disrupted nodes. Both in vivo and ex vivo, nodal proteins, such as Nav channels and ankyrin G, in intramuscular motor nerve bundles were completely lost by monoclonal antibody to GD1a ganglioside in association with complement deposition (McGonigal et al. 2010). Intriguingly, nodal Nav channel clusters were completely protected by both complement and calpain inhibition. Calpain, a calcium-dependent protease, is activated by the increased calcium influx through membrane pores formed by insertion of the membrane attack complex (MAC; a final product of the complement pathway) in the nodal axolemma. Nodal molecules including Nav channels, ankyrin G, or βIV spectrin, and neurofilament in the axonal cytoskeleton can be proteolyzed by calpain, and their breakdown can lead to nodal disruption and subsequent axon degeneration. Intraneural injection of monoclonal anti-GD1a antibody into rat sciatic nerves induced IgG and complement deposition at nodes and disrupted nodal and paranodal molecular organizations (Susuki et al. 2012). Importantly, electrophysiological analyses showed nerve conduction failure in both studies (McGonigal et al. 2010; Susuki et al. 2012), proving that the immune-mediated nodal disruption can indeed cause PNS dysfunction and thereby is responsible for development of neurological symptoms. Furthermore, the nerve conduction block induced by injecting anti-ganglioside IgG into rat sciatic nerve was rapidly resolved (Fig. 8.2c) (Susuki et al. 2012), identical to the reversible conduction failure in AMAN patients (Kuwabara et al. 1998; Kokubun et al. 2010). This type of nerve conduction block can be caused by multiple factors. For example, nodal ionic balance is impaired by the bidirectional, nonspecific ion and water pores formed by the insertion of the MAC into the nodal axolemma (McGonigal et al. 2010). This is clearly shown by drug treatments in the passive transfer model: complement inhibition protected nodal molecules and electrophysiological functions, whereas calpain inhibition (which does not remove the MAC) does not prevent nerve conduction failure despite preserved nodal Nav channel clusters (McGonigal et al. 2010). Furthermore, when nodal disruption progresses, the numbers of functioning Nav channels could decrease, driving current could leak from paranodal detachment, and juxtaparanodal Kv channels are exposed to the nodal area. All these factors could affect nerve conduction. Indeed, one study using a single rat myelinated nerve fiber preparation showed that anti-GM1 antibodies decrease the Na^+ current and cause a progressive increase of nonspecific leakage current in the presence of active complement (Takigawa et al. 1995), although another study showed differing results (Hirota et al. 1997).

8.5 Possible Nodal Disruption by Various Anti-ganglioside Antibodies

In addition to AMAN, other various conditions have been reported to be associated with anti-ganglioside antibodies. For example, serum IgM antibodies to GM1 gangliosides are frequently detected in multifocal motor neuropathy, a rare inflammatory neuropathy characterized by slowly progressive, asymmetrical distal limb weakness without sensory loss (Vlam et al. 2011). IgM anti-GM1 antibodies from patients bind to the nodes of Ranvier in rodent sciatic nerves but did not induce conduction failure (Harvey et al. 1995; Paparounas et al. 1999). However, another study showed that the intraneural injection of patient sera induced nerve conduction block in rat tibial nerves (Uncini et al. 1993). IgG antibodies against ganglioside GD1b disrupt nodes of Ranvier predominantly in sensory nerve fibers in rat sciatic nerves (Susuki et al. 2012), suggesting the underlying pathophysiology of rapidly reversible sensory nerve conduction failure in human acute sensory ataxic neuropathy associated with IgG anti-GD1b antibodies (Pan et al. 2001; Notturno et al. 2008). Rapidly reversible conduction failure has been reported in Fisher syndrome characterized by ophthalmoplegia, ataxia, and areflexia, and its related disorders associated with IgG antibodies against GQ1b ganglioside (Rajabally et al. 2011; Umapathi et al. 2012). Immune-mediated nodal disruption may also be involved in the pathophysiology of these conditions, although this is not yet proven because of the lack of human pathology or an experimental model (Uncini et al. 2013).

8.6 Nodal Disruption Caused by Immune-Mediated Demyelination

Myelinating Schwann cells can also be the direct target of autoimmune reactions. AIDP, the demyelinating form of GBS, is characterized pathologically by demyelination in the presence of lymphocytes and macrophages (Prineas 1981). Human pathology during the acute phase demonstrates the deposition of complement products along the outer surface of myelin and mild vesicular changes of the outermost myelin lamellae before the invasion of macrophages, suggesting that at least some forms of AIDP are complement-mediated attack on Schwann cells (Hafer-Macko et al. 1996b). Chronic inflammatory demyelinating polyradiculoneuropathy (CIDP) is characterized by limb weakness, sensory loss, and areflexia, and a relapsing or chronic progressive course, distinguishing the condition from GBS (Vallat et al. 2010; Pollard and Armati 2011). Segmental demyelination, remyelination, and lymphocytic infiltrates are present in the peripheral nerves from CIDP patients. The condition is responsive to several immunomodulative therapies such as intravenous immunoglobulin, plasma exchange, or corticosteroids. The detailed pathogenesis of AIDP or CIDP remains largely undiscovered.

Given the role of myelinating Schwann cells in organizing and maintaining nodes of Ranvier, it is not surprising that the defects in myelin can disrupt ion channel clustering at and near nodes. Indeed, skin biopsies from patients with demyelinating neuropathies including CIDP showed characteristic features, elongated nodes of Ranvier, dispersion of Caspr staining, and broadening of NF staining, providing a potential diagnostic marker of demyelinating neuropathies (Doppler et al. 2013). In superficial peroneal nerve biopsies from patients with CIDP, paranodal structures were altered; Caspr expression was upregulated and localized diffusely along the internodal segments (Cifuentes-Diaz et al. 2011). The nodal Nav channels were not altered but expression was also detected in the internodes. Intriguingly, a study on axonal membrane properties suggests that the decreased Nav channel densities at the node may be involved in the slowing of conduction in CIDP patients (Cappelen-Smith et al. 2001). Furthermore, some patients with CIDP show improvement of muscle strength over a few days following administration of intravenous immunoglobulin, suggesting reversible conduction block as a result of nodal/paranodal changes (Pollard and Armati 2011). Nodal/paranodal disruption has been further demonstrated in animal models for AIDP. Active immunization of rats with peripheral myelin induces an inflammatory demyelinating neuropathy, called experimental allergic neuritis (EAN), which is considered to be an animal model for AIDP. In spinal nerve roots from EAN animals, nodal Nav channel immunofluorescence changed from a highly focal ring to a more diffuse pattern, and as the disease progressed, it was eventually undetectable (Novakovic et al. 1998). In another EAN study, immunostaining of NF186 and gliomedin were often undetectable before demyelination (Lonigro and Devaux 2009). Taken together, these findings both in human pathology and in animal models strongly support the idea that disrupted paranodal junctions and altered localization of nodal Nav channels contribute to the pathophysiology of immune-mediated demyelinating neuropathies.

8.7 Nodal and Paranodal Proteins as Autoimmune Targets

Although the pathogenic roles of anti-ganglioside antibodies are well established in AMAN and related disorders, these antibodies are mostly negative in AIDP or CIDP patients (Willison and Yuki 2002; Pollard and Armati 2011). The autoimmune target is still unknown in the majority of AIDP and CIDP patients, although a small number of patients have antibodies against galactocerebroside, LM1 ganglioside, or myelin proteins such as P0 (Willison and Yuki 2002; Pollard and Armati 2011). Recent studies report that the autoantibodies against nodal or paranodal proteins gliomedin, NF186, NF155, contactin, or contactin/Caspr complex are detected in sera from patients with GBS or CIDP (Prüss et al. 2011; Devaux et al. 2012; Ng et al. 2012; Querol et al. 2013). Autoantibodies to NF186 and gliomedin are also found in the animal model, EAN (Lonigro and Devaux, 2009). Because these molecules are highly enriched at nodes or paranodes and are involved in the formation and stabilization of Nav channel clusters, these autoantibodies may disturb Nav channel function and destabilize nodal Nav channels similar to the pathophysiology seen in AMAN and eventually affect nerve conduction. Indeed, immunization against gliomedin induced a progressive neuropathy in Lewis rats characterized by conduction defects and nodal disruption in spinal nerve roots (Devaux 2012). Furthermore, passive administration of anti-gliomedin IgG into the EAN model (induced by immunization against the neuritogenic P2 peptide) augmented demyelination and nodal disruption and exacerbated the disease (Devaux 2012). Although the immunization of Lewis rats with NF186 did not induce significant neurological signs (Devaux 2012), administration of two different monoclonal antibodies against pan-NF (reactive to both NF186 and NF155) into the EAN model exacerbated the disease (Ng et al. 2012). These animal studies strongly suggest that the antibodies against nodal and paranodal proteins can contribute to the severity of AIDP or CIDP, although the pathogenic role of these autoantibodies in human patients remains to be tested (Hughes and Willison 2012). As these autoantibodies are detected in only a small portion of GBS or CIDP patients, and because some forms of AIDP are induced by complement-mediated autoimmune reactions on the outer surface of PNS myelin (Hafer-Macko et al. 1996b), the search for autoimmune targets of AIDP and CIDP will continue, including molecules located in the myelin sheath. Nevertheless, these findings suggest that the nodes and paranodes are potential targets of autoimmune reactions, and, consequently, immune-mediated nodal disruption widely underlies the pathophysiology of GBS and CIDP.

8.8 Conclusion

As reviewed here, dysfunction or disruption of the nodes of Ranvier has a significant role in the development of neurological symptoms in immune-mediated neuropathies. Autoantibodies against gangliosides cause AMAN via the complement pathway by disturbing the functions of Nav channels and destabilizing their clusters in peripheral motor nerve fibers. In AIDP and CIDP, nodes and paranodes are disrupted by demyelination, thereby damaging nerve transmission. Autoimmunity against proteins located at nodes and paranodes may also contribute to the pathophysiology of immune-mediated neuropathies. Finally, in addition to the immune-mediated neuropathies, dysfunction or disruption of nodes of Ranvier has been reported in various conditions including hereditary neuropathies, diabetic neuropathies, multiple sclerosis, and traumatic brain injury (Susuki 2013). Because axonal injury, demyelination, or both can disrupt the nodes of Ranvier, the concept of the dysfunction or disruption of nodes of Ranvier as a focus for understanding the pathophysiology can be widely applied to various neurological diseases and injuries.

Acknowledgments The author thanks Kelli L. Baalman and Dr. Matthew N. Rasband (Baylor College of Medicine, Houston, TX, USA) for comments on the manuscript.

References

Cappelen-Smith C, Kuwabara S, Lin CS, Mogyoros I, Burke D (2001) Membrane properties in chronic inflammatory demyelinating polyneuropathy. Brain 124:2439–2447

Cifuentes-Diaz C, Dubourg O, Irinopoulou T, Vigny M, Lachkar S, Decker L et al (2011) Nodes of Ranvier and paranodes in chronic acquired neuropathies. PLoS One 6:e14533

Devaux JJ (2012) Antibodies to gliomedin cause peripheral demyelinating neuropathy and the dismantling of the nodes of Ranvier. Am J Pathol 181:1402–1413

Devaux JJ, Odaka M, Yuki N (2012) Nodal proteins are target antigens in Guillain–Barré syndrome. J Peripher Nerv Syst 17:62–71
Doppler K, Werner C, Sommer C (2013) Disruption of nodal architecture in skin biopsies of patients with demyelinating neuropathies. J Peripher Nerv Syst 18:168–176
Eshed Y, Feinberg K, Poliak S, Sabanay H, Sarig-Nadir O, Spiegel I et al (2005) Gliomedin mediates Schwann cell–axon interaction and the molecular assembly of the nodes of Ranvier. Neuron 47:215–229
Feinberg K, Eshed-Eisenbach Y, Frechter S, Amor V, Salomon D, Sabanay H et al (2010) A glial signal consisting of gliomedin and NrCAM clusters axonal Na^+ channels during the formation of nodes of Ranvier. Neuron 65:490–502
Frenzen PD (2008) Economic cost of Guillain–Barré syndrome in the United States. Neurology 71:21–27
Griffin JW, Li CY, Macko C, Ho TW, Hsieh ST, Xue P et al (1996) Early nodal changes in the acute motor axonal neuropathy pattern of the Guillain–Barré syndrome. J Neurocytol 25:33–51
Hafer-Macko C, Hsieh ST, Li CY, Ho TW, Sheikh K, Cornblath DR et al (1996a) Acute motor axonal neuropathy: an antibody-mediated attack on axolemma. Ann Neurol 40:635–644
Hafer-Macko CE, Sheikh KA, Li CY, Ho TW, Cornblath DR, McKhann GM et al (1996b) Immune attack on the Schwann cell surface in acute inflammatory demyelinating polyneuropathy. Ann Neurol 39:625–635
Harvey GK, Toyka KV, Zielasek J, Kiefer R, Simonis C, Hartung HP (1995) Failure of anti-GM1 IgG or IgM to induce conduction block following intraneural transfer. Muscle Nerve 18:388–394
Hirota N, Kaji R, Bostock H, Shindo K, Kawasaki T, Mizutani K et al (1997) The physiological effect of anti-GM1 antibodies on saltatory conduction and transmembrane currents in single motor axons. Brain 120:2159–2169
Hughes RA, Willison HJ (2012) Neurofascin antibodies in inflammatory neuropathy: how many needles make a haystack? Neurology 79:2224–2225
Hughes RAC, Swan AV, Raphael JC, Annane D, van Koningsveld R, van Doorn PA (2007) Immunotherapy for Guillain–Barré syndrome: a systematic review. Brain 130:2245–2257
Kokubun N, Nishibayashi M, Uncini A, Odaka M, Hirata K, Yuki N (2010) Conduction block in acute motor axonal neuropathy. Brain 133:2897–2908
Kuwabara S, Yuki N, Koga M, Hattori T, Matsuura D, Miyake M et al (1998) IgG anti-GM1 antibody is associated with reversible conduction failure and axonal degeneration in Guillain–Barré syndrome. Ann Neurol 44:202–208
Lonigro A, Devaux JJ (2009) Disruption of neurofascin and gliomedin at nodes of Ranvier precedes demyelination in experimental allergic neuritis. Brain 132:260–273
McGonigal R, Rowan EG, Greenshields KN, Halstead SK, Humphreys PD, Rother RP et al (2010) Anti-GD1a antibodies activate complement and calpain to injure distal motor nodes of Ranvier in mice. Brain 133:1944–1960
Ng JK, Malotka J, Kawakami N, Derfuss T, Khademi M, Olsson T et al (2012) Neurofascin as a target for autoantibodies in peripheral neuropathies. Neurology 79:2241–2248
Notturno F, Caporale CM, Uncini A (2008) Acute sensory ataxic neuropathy with antibodies to GD1b and GQ1b gangliosides and prompt recovery. Muscle Nerve 37:265–268
Novakovic SD, Levinson SR, Schachner M, Shrager P (1998) Disruption and reorganization of sodium channels in experimental allergic neuritis. Muscle Nerve 21:1019–1032
Pan CL, Yuki N, Koga M, Chiang MC, Hsieh ST (2001) Acute sensory ataxic neuropathy associated with monospecific anti-GD1b IgG antibody. Neurology 57:1316–1318
Paparounas K, O'Hanlon GM, O'Leary CP, Rowan EG, Willison HJ (1999) Anti-ganglioside antibodies can bind peripheral nerve nodes of Ranvier and activate the complement cascade without inducing acute conduction block in vitro. Brain 122:807–816
Pereira JA, Lebrun-Julien F, Suter U (2012) Molecular mechanisms regulating myelination in the peripheral nervous system. Trends Neurosci 35:123–134

Poliak S, Peles E (2003) The local differentiation of myelinated axons at nodes of Ranvier. Nat Rev Neurosci 4:968–980

Pollard JD, Armati PJ (2011) CIDP – the relevance of recent advances in Schwann cell/axonal neurobiology. J Peripher Nerv Syst 16:15–23

Prineas JW (1981) Pathology of the Guillain–Barré syndrome. Ann Neurol 9(Suppl):6–19

Prüss H, Schwab JM, Derst C, Görtzen A, Veh RW (2011) Neurofascin as target of autoantibodies in Guillain–Barré syndrome. Brain 134:e173

Querol L, Nogales-Gadea G, Rojas-Garcia R, Martinez-Hernandez E, Diaz-Manera J, Suárez-Calvet X et al (2013) Antibodies to contactin-1 in chronic inflammatory demyelinating polyneuropathy. Ann Neurol 73:370–380

Rajabally YA, Hassan-Smith G, Notturno F, Eames PJ, Hayton T, Capasso M et al (2011) Motor and sensory conduction failure in overlap of Guillain–Barré and Miller Fisher syndrome: two simultaneous cases. J Neurol Sci 303:35–38

Salzer JL (2003) Polarized domains of myelinated axons. Neuron 40:297–318

Sejvar JJ, Baughman AL, Wise M, Morgan OW (2011) Population incidence of Guillain–Barré syndrome: a systematic review and meta-analysis. Neuroepidemiology 36:123–133

Sheikh KA, Sun J, Liu Y, Kawai H, Crawford TO, Proia RL et al (1999) Mice lacking complex gangliosides develop Wallerian degeneration and myelination defects. Proc Natl Acad Sci USA 96:7532–7537

Sherman DL, Brophy PJ (2005) Mechanisms of axon ensheathment and myelin growth. Nat Rev Neurosci 6:683–690

Susuki K, Nishimoto Y, Yamada M, Baba M, Ueda S, Hirata K et al (2003) Acute motor axonal neuropathy rabbit model: immune attack on nerve root axons. Ann Neurol 54:383–388

Susuki K, Baba H, Tohyama K, Kanai K, Kuwabara S, Hirata K et al (2007a) Gangliosides contribute to stability of paranodal junctions and ion channel clusters in myelinated nerve fibers. Glia 55:746–757

Susuki K, Rasband MN, Tohyama K, Koibuchi K, Okamoto S, Funakoshi K et al (2007b) Anti-GM1 antibodies cause complement-mediated disruption of sodium channel clusters in peripheral motor nerve fibers. J Neurosci 27:3956–3967

Susuki K, Yuki N, Schafer DP, Hirata K, Zhang G, Funakoshi K et al (2012) Dysfunction of nodes of Ranvier: a mechanism for anti-ganglioside antibody-mediated neuropathies. Exp Neurol 233:534–542

Susuki K (2013) Node of Ranvier disruption as a cause of neurological diseases. ASN Neuro 5:art:e00118. doi:10.1042/AN20130025

Susuki K, Chang K-J, Zollinger DR, Liu Y, Ogawa Y, Eshed-Eisenbach Y et al (2013) Three mechanisms assemble central nervous system nodes of Ranvier. Neuron 78:469–482

Takigawa T, Yasuda H, Kikkawa R, Shigeta Y, Saida T, Kitasato H (1995) Antibodies against GM1 ganglioside affect K^+ and Na^+ currents in isolated rat myelinated nerve fibers. Ann Neurol 37:436–442

Umapathi T, Tan EY, Kokubun N, Verma K, Yuki N (2012) Non-demyelinating, reversible conduction failure in Fisher syndrome and related disorders. J Neurol Neurosurg Psychiatry 83:941–948

Uncini A, Santoro M, Corbo M, Lugaresi A, Latov N (1993) Conduction abnormalities induced by sera of patients with multifocal motor neuropathy and anti-GM1 antibodies. Muscle Nerve 16:610–615

Uncini A, Susuki K, Yuki N (2013) Nodo-paranodopathy: beyond the demyelinating and axonal classification in anti-ganglioside antibody-mediated neuropathies. Clin Neurophysiol 124:1928–1934

Vallat JM, Sommer C, Magy L (2010) Chronic inflammatory demyelinating polyradiculoneuropathy: diagnostic and therapeutic challenges for a treatable condition. Lancet Neurol 9:402–412

Vlam L, van der Pol WL, Cats EA, Straver DC, Piepers S, Franssen H et al (2011) Multifocal motor neuropathy: diagnosis, pathogenesis and treatment strategies. Nat Rev Neurol 8:48–58

Willison HJ, Yuki N (2002) Peripheral neuropathies and anti-glycolipid antibodies. Brain 125:2591–2625

Yamashita T, Wu YP, Sandhoff R, Werth N, Mizukami H, Ellis JM et al (2005) Interruption of ganglioside synthesis produces central nervous system degeneration and altered axon–glial interactions. Proc Natl Acad Sci USA 102:2725–2730

Yuki N, Hartung HP (2012) Guillain–Barré syndrome. N Engl J Med 366:2294–2304

Yuki N, Yamada M, Koga M, Odaka M, Susuki K, Tagawa Y et al (2001) Animal model of axonal Guillain–Barré syndrome induced by sensitization with GM1 ganglioside. Ann Neurol 49: 712–720

Chapter 9
Pathogenesis of Diabetic Neuropathy from the Point of View of Schwann Cell Abnormalities

Koichi Kato, Eva L. Feldman, and Jiro Nakamura

Abstract Various factors have been implicated in the pathogenesis of diabetic neuropathy, such as polyol pathway hyperactivity, abnormal protein kinase C activity, increased oxidative stress, increased nonenzymatic glycation, and reduced synthesis of neurotrophic factors. C-peptide deficiency is also involved in the pathogenesis of diabetic neuropathy in type 1 diabetes. These glucose-mediated metabolic abnormalities affect all cellular components of nerve tissue including Schwann cells, neurons, and endoneurial endothelial cells. In diabetes, Schwann cells themselves undergo hyperglycemia insults, and the supporting functions of Schwann cells for neurons are also disturbed by high glucose, resulting in dysfunction of both neurons and Schwann cells.

Among the treatments that are based on pathogenic mechanisms of diabetic neuropathy, medications such as α-lipoic acid, benfotiamine, actovegin, and epalrestat are available for clinical use. However, the efficacy of these agents is limited and unsatisfactory.

The precise mechanism of action of each pathogenic factor and the interaction of these factors remain unknown. Therefore, the pathophysiological mechanisms should be elucidated for all cellular components of nerve tissue, including Schwann cells. Elucidation of the pathogenesis of diabetic neuropathy is essential for establishing effective treatment for this neuropathy.

K. Kato (✉)
Laboratory of Medicine, Aichi Gakuin University School of Pharmacy,
1-100 Kusumoto-cho, Chikusa-ku, Nagoya 464-8650, Japan
e-mail: kkato@dpc.agu.ac.jp

E.L. Feldman
Department of Neurology, University of Michigan, Ann Arbor, MI, USA

J. Nakamura
Division of Diabetes, Department of Internal Medicine, Aichi Medical University
School of Medicine, Nagakute, Aichi, Japan

K. Sango and J. Yamauchi (eds.), *Schwann Cell Development and Pathology*,
DOI 10.1007/978-4-431-54764-8_9,

Keywords C-peptide • Diabetic neuropathy • Glycation • Neurotrophin secretion • Oxidative stress • Pathogenesis • Polyol pathway • Protein kinase C • Schwann cells

9.1 Introduction

Diabetic peripheral neuropathy is one of the most common microvascular complications of diabetes, eventually affecting at least 50 % of patients with both type 1 and type 2 diabetes (Little et al. 2007). Diabetic neuropathy is characterized by progressive, nerve length-dependent loss of peripheral nerve fibers in a symmetrical "stocking and glove" manner (Said 2007), causing decreased sensation, spontaneous pain, and eventually complete loss of sensation. It is also the leading cause of nontraumatic amputations.

Various factors have been implicated in the pathogenesis of diabetic neuropathy such as polyol pathway hyperactivity, abnormal protein kinase C (PKC) activity, increased oxidative stress, increased nonenzymatic glycation, abnormal hexosamine pathways, and reduced synthesis of neurotrophic factors (Fig. 9.1). (Obrosova 2009). C-peptide deficiency is also involved in the pathogenesis of diabetic

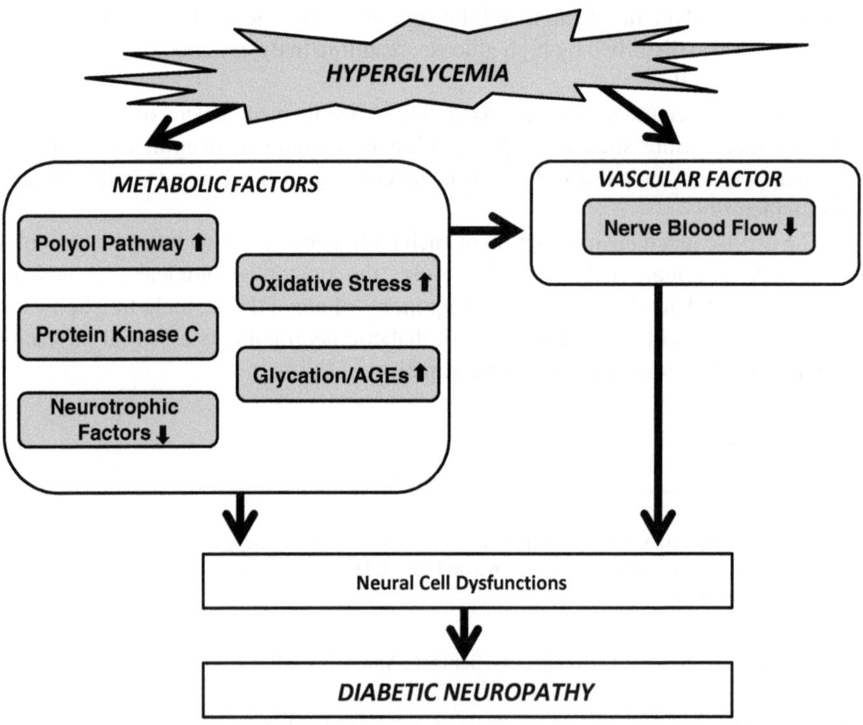

Fig. 9.1 Pathogenesis of diabetic neuropathy

neuropathy in type 1 diabetes (Kato et al. 2003; unpublished data). Vascular factors such as decreased nerve blood flow and increased aggregation, as well as the metabolic factors just described, are also considered important in the pathogenesis (Nakamura et al. 1999). All these glucose-mediated metabolic and vascular abnormalities have a close relationship and interaction with one another and affect the peripheral nerve tissues.

Peripheral nerve tissues consist of three cellular components important in the pathophysiology of diabetic neuropathy: neurons, Schwann cells, and endothelial cells of endoneurial vessels. Axons extend from neurons, and Schwann cells wrap themselves around the axons, forming myelin sheaths. Endothelial cells of endoneurial vessels supply nerve blood flow and nutrients to nerve tissues. An understanding of diabetic neuropathy requires elucidation of the defects and the interactions of these three kinds of cells under diabetic conditions.

Schwann cells are glial cells of the peripheral nerve system whose function is to support neurons. As described in Chaps. 2–5, Schwann cells ensheath their axons in myelinating or unmyelinating forms. These cells are involved in a variety of important aspects of peripheral nerve biology, including indispensable conduction, nerve development and regeneration, neurotrophic factor support for neurons, and production of the nerve extracellular matrix.

In diabetes, the Schwann cells themselves undergo hyperglycemic insults, and the supporting functions of Schwann cells for neurons are also disturbed by high glucose, resulting in dysfunction of both neurons and Schwann cells. These dysfunctions result in defects of nerve function such as decreased nerve conduction velocity, decreased neurotrophic factor secretion, and diminished Na^+/K^+-ATPase activity. It has been reported that disrupted Schwann cell mitochondrial function in connection with perturbed glial support causes primary neuronal degeneration, first in unmyelinated fibers and subsequently in myelinated fibers (Viader et al. 2011).

Among the treatments that are based on pathogenic mechanisms of diabetic neuropathy, antioxidant α-lipoic acid, benfotiamine (a vitamin B_1 derivative), actovegin, and the aldose reductase (AR) inhibitor epalrestat are available for clinical use (Zenker et al. 2013). However, the efficacy of these agents is limited and unsatisfactory in improving severe diabetic neuropathy. Development of an effective treatment is strongly desired.

In this chapter, the pathogenesis of diabetic neuropathy, focusing particularly on Schwann cells, is discussed.

9.2 IMS32 Cells for the Study of Diabetic Neuropathy

Immortalized mouse Schwann cells (IMS32 cells) were established by isolating Schwann cells from mouse dorsal root ganglion (DRG) and peripheral nerves (Watabe et al. 1995). As described in Chap. 10, these immortalized cells preserve well the characteristics of Schwann cells and are a suitable model for studying diabetic neuropathy (Sango et al. 2011). Most research on Schwann cells has been performed using malignant cell lines such as JS1 schwannoma cells (Kamiya et al. 2003) or primary culture of Schwann cells. These immortalized cells are useful for studying diabetic neuropathy for the following reasons:

1. High glucose increases polyol pathway flux (Sango et al. 2006)
2. High glucose decreases cell proliferation (Kato et al. 2003; unpublished data; see Sect. 9.3.6)
3. IMS32 cells respond to several cytokine and neurotrophic factors (Watabe et al. 1995)
4. Cell cultures of IMS32 cells are simple and do not require any special agent such as neuregulin (NRG)-β and forskolin, which is necessary for primary cultures of Schwann cells isolated from DRG

Because of these advantages, IMS32 cells have been employed for in vitro studies of diabetic neuropathy (Kato et al. 2003; Ota et al. 2007; Tosaki et al. 2008; Sango et al. 2008; Suzuki et al. 2011; Kim et al. 2013).

9.3 Pathogenesis of Diabetic Neuropathy

9.3.1 Polyol Pathway Hyperactivity

The polyol pathway is a two-step metabolic pathway in which glucose is reduced to sorbitol, which is then oxidized to fructose (Gabbay 1973). In the first and rate-limiting step of this pathway, glucose is metabolized to sorbitol by NADPH-dependent AR. Sorbitol is then converted to fructose by NAD-dependent sorbitol dehydrogenase. At physiological glucose concentrations, sorbitol synthesis via the polyol pathway is less than 3 %; however, in diabetes, glucose flux into this pathway is increased, and 30–35 % of the glucose can be converted to sorbitol. Subsequent accumulation of sorbitol and fructose occurs with concomitant myo-inositol depletion in neural cells including Schwann cells and neurons. This myo-inositol depletion causes a decrease in phosphoinositide and diacylglycerol levels, resulting in diminished PKC activity; in turn this reduces Na^{+}/K^{+}-ATPase activity, leading to a deficit in nerve conduction velocity and nerve dysfunction (Greene et al. 1987; Nakamura et al. 1992; Kato et al. 1999).

In JS1 schwanomma cells, high glucose inhibited Schwann cell proliferation with the accumulation of sorbitol and fructose and depletion of myo-inositol (Kamiya et al. 2003). Epalrestat, an AR inhibitor, ameliorated these abnormalities, suggesting that these deficits are mediated, at least in part, by polyol pathway hyperactivity in JS1 Schwannoma cells. In IMS32 cells, high glucose also caused an accumulation of sorbitol and fructose that was reversed by AR inhibitors. In IMS32 cells, AR gene expression was also induced by high glucose and methylglyoxal, indicating that AR induction by high glucose may lead to further intensification of polyol pathway hyperactivity (Sango et al. 2006, 2008).

9.3.2 Protein Kinase C Activity Abnormality

Tissue- and cell-specific differences exist, depending on the type of diabetic complication (Ishii et al. 1996; Koya and King 1998), in hyperglycemia-induced PKC activity abnormalities in the pathogenesis of diabetic complications. In smooth muscle cells (Inoguchi et al. 1992, 1994; Yasuda et al. 2001; Nakamura et al. 2001), endothelial cells, and mesangial cells (Koya et al. 1997), an increase in PKC activity, particularly that of PKC-β, has been reported, indicating that an increase in PKC-β activation plays an important role in the pathogenesis of diabetic macroangiopathy, retinopathy, and nephropathy. On the other hand, in JS-1 schwannoma cells, in contrast to smooth muscle cells and mesangial cells, a decrease in PKC activity, particularly PKC-α activity, has been reported with reduced cell proliferation, and these deficits are mediated via polyol pathway hyperactivity, as described in the section on polyol pathway in this chapter (Kamiya et al. 2003).

Another important component of diabetic neuropathy is endoneurial endothelial cells (Cameron et al. 2001). PKC-β activation in the endoneurial microvasculature can cause a reduction of nerve blood flow. A PKC-β-specific inhibitor, LY333531, has been reported to prevent a delay in motor nerve conduction velocity and a decrease in sciatic nerve blood flow in diabetic rats despite unchanged PKC activity in whole nerve tissue; thus, the beneficial effect of PKC-β inhibitor on this neuropathy may be a result of ameliorating the decreased nerve blood flow in diabetic rats (Nakamura et al. 1999).

Thus, the cellular components of nerve tissue exhibit cell-specific differences in PKC activity abnormalities that play important roles in the pathogenesis of diabetic neuropathy.

9.3.3 Oxidative Stress

A considerable number of studies indicate that glucose-induced oxidative stress is a key factor in the development of diabetic neuropathy (Vincent et al. 2011).

High glucose increases oxidative stress, measured using oxidative stress markers such as 4-hydroxy-2-nonenal (4HNE), acrolein (ACR), and hexanoyl lysine (HEL) in IMS32 cells (Sango et al. 2008) and nitrated proteins in human Schwann cells (Askwith et al. 2012). Although neurons or DRGs undergo apoptosis under high-glucose conditions by glucose-increased oxidative stress in mitochondria (Vincent et al. 2009), apoptosis did not occur in IMS32 cells or human Schwann cells as a result of high glucose alone in in vitro studies. Although a glucose-dependent increase in superoxide production occurs in embryonic sensory neurons, hyperglycemia did not induce a substantial change in superoxide levels in Schwann cells (Zhang et al. 2010). Other potent insults such as methylglyoxal, a very strong inducer of oxidative stress, or palmitic acid, which causes lipotoxity, may be capable of inducing apoptosis in Schwann cells (Fukunaga et al. 2004; Ota et al. 2007).

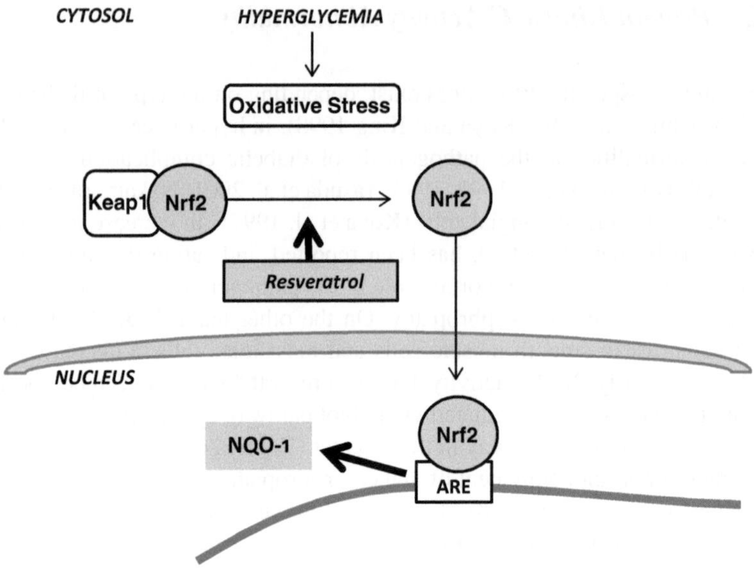

Fig. 9.2 Regulatory mechanisms of antioxidant enzyme NAD(P)H:quinone oxidoreductase-1 (NQO1) induction by resveratrol via translocation of Nrf2 from the cytosol into the nuclei

Decreased cell proliferation and migration have been observed in human Schwann cells (Askwith et al. 2012), Schwann cells of primary cultures isolated from mouse (Gumy et al. 2008), and IMS32 cells (Kato et al. 2003). This decrease in cell proliferation may restrict the regeneration of axons and Schwann cells (Gumy et al. 2008), an important aspect of diabetic neuropathy. The fact that α-lipoic acid and epalrestat restored the decrease in proliferation under high-glucose conditions indicates that this deficit of cell proliferation is modulated, at least in part, by oxidative stress or polyol pathways (Askwith et al. 2012; Kamiya et al. 2003).

Enhancing antioxidant mechanisms in neural cells is a unique target of therapeutics. Resveratrol, a polyphenol present in grape skin, has potent antioxidant and antiinflammatory effects and has been reported to prevent not only aging and cancer but also diabetic nephropathy (Jang et al. 1997; Baur et al. 2006; Kitada et al. 2011). We observed the effect of resveratrol and glucose on the expression of NAD(P)H:quinone oxidoreductase-1 (NQO1), an antioxidant enzyme, in Schwann cells and DRG and investigated the mechanisms involved (Fig. 9.2) (Vincent et al. 2009; Kato et al. 2005; unpublished data). As shown in Fig. 9.3, resveratrol induced NQO1 expression in Schwann cells whereas high glucose did not alter NQO1 expression. These results suggest that the antioxidant mechanism is enhanced by resveratrol and that hyperglycemia is not a potent inducer of antioxidant enzymes such as NQO1. Because antioxidant enzymes are known to be induced by the translocation of the transcriptional factor Nrf2 into nuclei, we investigated Nrf2 protein expression in a nucleus fraction separated by ultracentrifugation in Schwann cells. Both resveratrol and high glucose increased Nrf2 translocation into the nucleus,

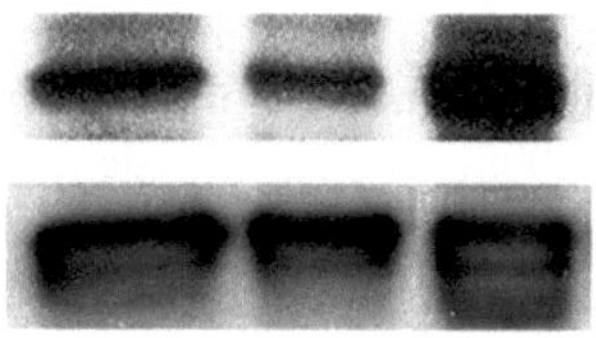

Fig. 9.3 Resveratrol (RES) induced the expression of the antioxidant enzyme NAD(P)H:quinone oxidoreductase-1 (NQO1) in Schwann cells. In contrast, high glucose (Glu) did not alter NQO1 protein expression compared with normal glucose (Cont)

Fig. 9.4 Both high glucose (Glu) and resveratrol (RES) induced translocation of Nrf2 into the nucleus in Schwann cells. Resveratrol induced Nrf2 translocation to a greater extent than high glucose in Schwann cells

with resveratrol inducing Nrf2 translocation to greater extent than high glucose (Fig. 9.4). The fact that high glucose increased Nrf2 translocation, but not NQO1 expression, suggests that high glucose induces oxidative stress-initiated translocation of Nrf2 into the nucleus. However, the degree of Nrf2 translocation by high glucose may not be enough for induction of NQO-1 expressions. These results indicate that resveratrol may exert beneficial effects on diabetic neuropathy by enhancing antioxidant defense mechanisms in Schwann cells.

9.3.4 Glycation

Nonenzymatic glycation and excessive formation of advanced glycation end products (AGEs) are implicated in the pathogenesis of diabetic neuropathy. AGEs are known to accumulate in the peripheral nerves of patients with diabetes (Sugimoto et al. 2008). Various types of AGEs are formed and increased in the serum of patients with diabetes. Among them, AGEs derived from glyceraldehyde and glycoaldehyde, but not those derived from glucose, induced apoptosis mediated by p38 mitogen-activated protein (MAP) kinase in Schwann cells and also led to the activation of activated nuclear factor-kappa B (NF-kappa B) and proinflammatory cytokines tumor necrosis factor (TNF)-α and interleukin (IL)-1β (Sekido et al. 2004). As described in the section on oxidative stress, methylglyoxal is a potent precursor of AGEs and, similar to AGEs, induces apoptosis in Schwann cells via p38 MAPK (Fukunaga et al. 2004). Metformin, an oral hypoglycemic biguanide, prevented MG-induced apoptotic signals in IMS32 cells by inhibiting the formation of AGEs and reactive oxygen species (Ota et al. 2007).

Receptor for AGEs (RAGE) is also involved in diabetic neuropathy. Binding of ligands to RAGE results in activation of NF-kappa B and subsequent secretion of cytokines (Toth et al. 2008). In STZ diabetic rats, diabetic sural axons, Schwann cells, and sensory neurons in ganglia showed marked and cumulative increases in RAGE mRNA and protein levels. RAGE-mediated signaling pathway activation for NF-kappa B and PKC-βII pathways was the most evident among Schwann cells in DRG and peripheral nerves. Furthermore, RAGE expressions in Schwann cells were also increased in sural nerve biopsies from patients with diabetic polyneuropathy (Haslbeck et al. 2007).

9.3.5 Impaired Neurotrophin Secretion

Impaired neurotrophic support plays an important role in diabetic neuropathy. Several neurotrophic factors including insulin, insulin-like growth factor (IGF)-1, nerve growth factor (NGF), ciliary neurotrophic factor, and neurotrophin-3 (NT-3) have been reported to have a beneficial effect on diabetic neuropathy (Christianson et al. 2003).

NT-3-induced NGF production was significantly suppressed when Schwann cells were cultured under high-glucose conditions for 5 weeks (Suzuki et al. 2004). In these cells, the levels of glutathione (GSH) and cAMP response element-binding protein (CREB) were reduced, whereas the level of NF-kappa B was elevated. These changes were abolished by the AR inhibitor fidarestat, suggesting a polyol pathway-mediated mechanism. Conditioned medium collected from IMS32 cells cultured under high glucose for 4 days exhibited decreased NGF concentrations compared with that collected from cells cultured under normal glucose (Tosaki et al. 2008), and the neurite outgrowth cultured with IMS media obtained under high-glucose conditions was significantly reduced compared with that obtained under normal conditions, suggesting that reduced NGF secretion from Schwann cells may cause a defect of axonal regeneration. In contrast, it has been reported that short-time incubation of Schwann cells with high glucose for 24 h increased NGF and brain-derived neurotrophic factor (BDNF) production (Zhu et al. 2012). Given that diabetic neuropathy develops over years, short-time incubation with high glucose may not reflect the pathophysiology of diabetic neuropathy.

Neuregulins, growth factors that bind to Erb receptor tyrosine kinases in Schwann cells, are known to promote cell survival, mitogenesis, and myelination in undifferentiated Schwann cells, and also to induce demyelination of myelinated Schwann cells. A recent study revealed the role of hyperglycemia-induced increase in Erb B2 activity and a concomitant decrease in the protein caveolin-1 in NRG-induced Schwann cell demyelination (Yu et al. 2008). A study using the Cav1 knockout mouse demonstrated that altered Erb B2 signaling is a novel mechanism that contributes to Schwann cell dysfunction in diabetes and that inhibiting Erb B2 may be a therapeutic approach for diabetic neuropathy (McGuire et al. 2009).

9.3.6 C-Peptide

C-peptide, a by-product of insulin, is well known as a marker of insulin secretion in diabetes. It has been found to have various biological and physiological activities including insulin-mimetic effects (Zierath et al. 1991). Furthermore, C-peptide ameliorated diabetic neuropathy in patients with type 1 diabetes (Ekberg et al. 2003; Ido et al. 1997; Sima et al. 2001). Given that the precise mechanisms of C-peptide on diabetic neuropathy remain unclear, we investigated the effects of high glucose and C-peptide on proliferation and MAP kinase activities in IMS32 cells (Kato et al. 2003; unpublished data). IMS32 cells were cultured with 5.5 or 20 mM glucose in the presence or absence of C-peptide. Compared with control cells, IMS cells cultured with 20 mM glucose for 28 days demonstrated significantly decreased thymidine incorporation. This decrease was almost normalized by 0.3 nM C-peptide, which is the physiological concentration of C-peptide. Because MAP kinase has an important role in cell proliferation, total and phosphorylated-ERK1/2 expressions were measured. Compared with control cells, short-time incubation with 0.3 nM C-peptide for 20 min increased ERK1/2 activities approximately twofold.

These observations suggest that beneficial effects of C-peptide on type 1 diabetic neuropathy may be mediated, at least in part, by amelioration of Schwann cell growth via activation of MAP kinase. Thus, C-peptide can be a target for novel therapeutics for diabetic neuropathy.

9.4 Conclusion

As described here, a variety of pathogenic factors are involved in the pathogenesis of diabetic neuropathy. However, the precise mechanism of action of each pathogenic factor and the interaction of these factors remain unknown. Therefore, the pathophysiological mechanisms should be determined for all cellular components of nerve tissue such as Schwann cells, neurons, and endoneurial endothelial cells. To this end, in vitro experiments using neuronal cells such as immortalized Schwann cells are useful tools for investigating the pathogenic mechanisms.

Considering that blockade of one pathogenic factor will not be sufficient to improve diabetic neuropathy, a multifactorial intervention that ameliorates multiple major pathogenic factors at the same time, along with strict treatment for diabetes mellitus, hypertension, and dyslipidemia, may be required.

Thus, elucidating the pathogenesis of diabetic neuropathy is essential for the establishment of effective treatment for this neuropathy.

Acknowledgments I thank Dr. Kazunori Sango for helpful suggestions and Enago (www.enago.jp) for English language review.

References

Askwith T, Zeng W, Eggo MC, Stevens MJ (2012) Taurine reduces nitrosative stress and nitric oxide synthase expression in high glucose-exposed human Schwann cells. Exp Neurol 233(1):154–162

Baur JA, Pearson KJ, Price NL, Jamieson HA, Lerin C, Kalra A, Prabhu VV, Allard JS, Lopez-Lluch G, Lewis K, Pistell PJ, Poosala S, Becker KG, Boss O, Gwinn D, Wang M, Ramaswamy S, Fishbein KW, Spencer RG, Lakatta EG, Le Couteur D, Shaw RJ, Navas P, Puigserver P, Ingram DK, de Cabo R, Sinclair DA (2006) Resveratrol improves health and survival of mice on a high-calorie diet. Nature (Lond) 444:337–342

Cameron NE, Eaton SE, Cotter MA, Tesfaye S (2001) Vascular factors and metabolic interactions in the pathogenesis of diabetic neuropathy. Diabetologia 44(11):1973–1988

Christianson JA, Ryals JM, McCarson KE, Wright DE (2003) Beneficial actions of neurotrophin treatment on diabetes-induced hypoalgesia in mice. J Pain 4:493–504

Ekberg K, Brismar T, Johansson BL, Jonsson B, Lindström P, Wahren J (2003) Amelioration of sensory nerve dysfunction by C-peptide in patients with type 1 diabetes. Diabetes 52:536–541

Fukunaga M, Miyata S, Liu BF, Miyazaki H, Hirota Y, Higo S, Hamada Y, Ueyama S, Kasuga M (2004) Methylglyoxal induces apoptosis through activation of p38 MAPK in rat Schwann cells. Biochem Biophys Res Commun 320(3):689–695

Gabbay KH (1973) The sorbitol pathway and the complications of diabetes. N Engl J Med 288:831–836

Greene DA, Lattimer SA, Sima AA (1987) Sorbitol, phosphoinositides, and sodium-potassium-ATPase in the pathogenesis of diabetic complications. N Engl J Med 316(10):599–606

Gumy LF, Bampton ET, Tolkovsky AM (2008) Hyperglycaemia inhibits Schwann cell proliferation and migration and restricts regeneration of axons and Schwann cells from adult murine DRG. Mol Cell Neurosci 37(2):298–311

Haslbeck KM, Neundörfer B, Schlötzer-Schrehardtt U, Bierhaus A, Schleicher E, Pauli E, Haslbeck M, Hecht M, Nawroth P, Heuss D (2007) Activation of the RAGE pathway: a general mechanism in the pathogenesis of polyneuropathies? Neurol Res 29(1):103–110

Ido Y, Vindigni A, Chang K, Stramm L, Chance R, Heath WF, DiMarchi RD, Di Cera E, Williamson JR (1997) Prevention of vascular and neural dysfunction in diabetic rats by C-peptide. Science 277:563–566

Inoguchi T, Battan R, Handler E, Sportsman JR, Heath W, King GL (1992) Preferential elevation of protein kinase C isoform βII and diacylglycerol levels in the aorta and heart of diabetic rats: differential reversibility to glycemic control by islet cell transplantation. Proc Natl Acad Sci USA 89:11059–11063

Inoguchi T, Xia P, Higashi S, Kunisaki M, Feener EP, King GL (1994) Insulin's effect on protein kinase C and diacylglycerol induced by diabetes and glucose in vascular tissues. Am J Physiol 267:E369–E379

Ishii H, Jirousek MR, Koya D, Takagi C, Xia P, Clermont A, Bursell SE, Kern TS, Ballas LM, Heath WF, Stramm LE, Feener EP, King GL (1996) Amelioration of vascular dysfunction in diabetic rats by an oral PKC β inhibitor. Science 272:728–731

Jang M, Cai L, Udeani GO, Slowing KV, Thomas CF, Beecher CW, Fong HH, Farnsworth NR, Kinghorn AD, Mehta RG, Moon RC, Pezzuto JM (1997) Cancer chemopreventive activity of resveratrol, a natural product derived from grapes. Science 275:218–220

Kamiya H, Nakamura J, Hamada Y, Nakashima E, Naruse K, Kato K, Yasuda Y, Hotta N (2003) Polyol pathway and protein kinase C activity of rat schwannoma cells. Diabetes Metab Res Rev 19(2):131–139

Kato K, Thomas TP, Stevens MJ, Greene DA, Nakamura J (1999) 2-Chloroadenosine reverses hyperglycemia-induced inhibition of phosphoinositide synthesis in cultured human retinal pigment epithelial cells and prevents reduced nerve conduction velocity in diabetic rats. Metabolism 48(7):827–833

Kato K, Nakamura J, Kamiya H, Naruse K, Nakashima E, Hamada Y, Sakakibara F, Kusunoki M, Watabe K, Otake K (2003) Effect of high glucose and C-peptide on proliferation and MAP kinase activity in cultured immortalized mouse Schwann (IMS32) cells. In: Abstracts of the 6th diabetic neuropathy satellite symposium of the 18th international diabetes federation congress in conjunction with 13th annual meeting of the diabetic neuropathy study group of EASD (Neurodiab XII), Saint Malo, 30 Aug–2 Sept 2003

Kato K, Vincent A, Buccus C, Feldman E (2005) A novel antioxidant, resveratrol, prevent high glucose-induced apoptosis via induction of NAD(P)H:quinone oxidoreductase-1 (NQO1) in rat dorsal root ganglion. In: Abstracts of American Neurological Association 130th annual meeting, San Diego, 25–28 Sept 2005

Kim ES, Isoda F, Kurland I, Mobbs CV (2013) Glucose-induced metabolic memory in Schwann cells: prevention by PPAR agonists. Endocrinology 154(9):3054–3066

Kitada M, Kume S, Imaizumi N, Koya D (2011) Resveratrol improves oxidative stress and protects against diabetic nephropathy through normalization of Mn-SOD dysfunction in AMPK/SIRT1-independent pathway. Diabetes 60(2):634–643

Koya D, King GL (1998) Protein kinase C activation and the development of diabetic complications. Diabetes 47:859–866

Koya D, Jirousek MR, Lin Y-W, Ishii H, Kuboki K, King GL (1997) Characterization of protein kinase Cβ isoform activation on the gene expression of transforming growth factor-β extracellular matrix components, and prostanoids in the glomeruli of diabetic rats. J Clin Invest 100:115–126

Little AA, Edwards JL, Feldman EL (2007) Diabetic neuropathies. Pract Neurol 7:82–92

McGuire JF, Rouen S, Siegfreid E, Wright DE, Dobrowsky RT (2009) Caveolin-1 and altered neuregulin signaling contribute to the pathophysiological progression of diabetic peripheral neuropathy. Diabetes 58(11):2677–2686

Nakamura J, Del Monte MA, Shewach D, Lattimer SA, Greene DA (1992) Inhibition of phosphatidylinositol synthase by glucose in human retinal pigment epithelial cells. Am J Physiol 262 (4 pt 1):E417–E426

Nakamura J, Kato K, Hamada Y, Nakayama M, Chaya S, Nakashima E, Naruse K, Kasuya Y, Mizubayashi R, Miwa K, Yasuda Y, Kamiya H, Ienaga K, Sakakibara F, Koh N, Hotta N (1999) A protein kinase C-β-selective inhibitor ameliorates neural dysfunction in streptozotosin-induced diabetic rats. Diabetes 48:2090–2095

Nakamura J, Kasuya Y, Hamada Y, Nakashima E, Naruse K, Yasuda Y, Kato K, Hotta N (2001) Glucose-induced hyperproliferation of cultured rat aortic smooth muscle cells through polyol pathway hyperactivity. Diabetologia 44:480–487

Obrosova IG (2009) Diabetes and the peripheral nerve. Biochim Biophys Acta 1792(10): 931–940

Ota K, Nakamura J, Li W, Kozakae M, Watarai A, Nakamura N, Yasuda Y, Nakashima E, Naruse K, Watabe K, Kato K, Oiso Y, Hamada Y (2007) Metformin prevents methylglyoxal-induced apoptosis of mouse Schwann cells. Biochem Biophys Res Commun 357(1):270–275

Said G (2007) Diabetic neuropathy: a review. Nat Clin Pract Neurol 3:331–340

Sango K, Suzuki T, Yanagisawa H, Takaku S, Hirooka H, Tamura M, Watabe K (2006) High glucose-induced activation of the polyol pathway and changes of gene expression profiles in immortalized adult mouse Schwann cells IMS32. J Neurochem 98(2):446–458

Sango K, Yanagisawa H, Kato K, Kato N, Hirooka H, Watabe K (2008) Differential effects of high glucose and methylglyoxal on viability and polyol metabolism in immortalized adult mouse Schwann cells. Open Diabetes J 1:1–11

Sango K, Yanagisawa H, Takaku S, Kawakami E, Watabe K (2011) Immortalized adult rodent Schwann cells as in vitro models to study diabetic neuropathy. Exp Diabetes Res 2011:374943

Sekido H, Suzuki T, Jomori T, Takeuchi M, Yabe-Nishimura C, Yagihashi S (2004) Reduced cell replication and induction of apoptosis by advanced glycation end products in rat Schwann cells. Biochem Biophys Res Commun 320(1):241–248

Sima AA, Zhang W, Sugimoto K, Henry D, Li Z, Wahren J, Grunberger G (2001) C-Peptide prevents and improves chronic type I diabetic polyneuropathy in the BB/Wor rat. Diabetologia 44:889–897

Sugimoto K, Yasujima M, Yagihashi S (2008) Role of advanced glycation end products in diabetic neuropathy. Curr Pharm Des 14(10):953–961

Suzuki T, Sekido H, Kato N, Nakayama Y, Yabe-Nishimura C (2004) Neurotrophin-3-induced production of nerve growth factor is suppressed in Schwann cells exposed to high glucose: involvement of the polyol pathway. J Neurochem 91(6):1430–1438

Suzuki J, Akahane K, Nakamura J, Naruse K, Kamiya H, Himeno T, Nakamura N, Shibata T, Kondo M, Nagasaki H, Fujiya A, Oiso Y, Hamada Y (2011) Palmitate induces apoptosis in Schwann cells via both ceramide-dependent and independent pathways. Neuroscience 176:188–198

Tosaki T, Kamiya H, Yasuda Y, Naruse K, Kato K, Kozakae M, Nakamura N, Shibata T, Hamada Y, Nakashima E, Oiso Y, Nakamura J (2008) Reduced NGF secretion by Schwann cells under the high glucose condition decreases neurite outgrowth of DRG neurons. Exp Neurol 213(2):381–387

Toth C, Rong LL, Yang C, Martinez J, Song F, Ramji N, Brussee V, Liu W, Durand J, Nguyen MD, Schmidt AM, Zochodne DW (2008) Receptor for advanced glycation end products (RAGEs) and experimental diabetic neuropathy. Diabetes 57(4):1002–1017

Viader A, Golden JP, Baloh RH, Schmidt RE, Hunter DA, Milbrandt J (2011) Schwann cell mitochondrial metabolism supports long-term axonal survival and peripheral nerve function. J Neurosci 31:10128–10140

Vincent AM, Kato K, McLean LL, Soules ME, Feldman EL (2009) Sensory neurons and Schwann cells respond to oxidative stress by increasing antioxidant defense mechanisms. Antioxid Redox Signal 11(3):425–438

Vincent AM, Callaghan BC, Smith AL, Feldman EL (2011) Diabetic neuropathy: cellular mechanisms as therapeutic targets. Nat Rev Neurol 7(10):573–583

Watabe K, Fukuda T, Tanaka J, Honda H, Toyohara K, Sakai O (1995) Spontaneously immortalized adult mouse Schwann cells secrete autocrine and paracrine growth-promoting activities. J Neurosci Res 41(2):279–290

Yasuda Y, Nakamura J, Hamada Y, Nakayama M, Chaya S, Naruse K, Nakashima E, Kato K, Kamiya H, Hotta N (2001) Role of PKC and TGF-β receptor in glucose-induced proliferation of smooth muscle cells. Biochem Biophys Res Commun 281:71–77

Yu C, Rouen S, Dobrowsky RT (2008) Hyperglycemia and downregulation of caveolin-1 enhance neuregulin-induced demyelination. Glia 56:877–887

Zenker J, Ziegler D, Chrast R (2013) Novel pathogenic pathways in diabetic neuropathy. Trends Neurosci 36(8):439–449. doi:10.1016/j.tins.04.008

Zhang L, Yu C, Vasquez FE, Galeva N, Onyango I, Swerdlow RH, Dobrowsky RT (2010) Hyperglycemia alters the Schwann cell mitochondrial proteome and decreases coupled respiration in the absence of superoxide production. J Proteome Res 9(1):458–471

Zhu H, Yu WJ, Le Y, Wang WJ, Li F, Gui T, Wang YM, Shi WD, Ding WL, Fan XQ (2012) High glucose levels increase the expression of neurotrophic factors associated with p-p42/p44 MAPK in Schwann cells in vitro. Mol Med Rep 6(1):179–184

Zierath JR, Galuska D, Johansson BL, Wallberg-Henriksson H (1991) Effect of human C-peptide on glucose transport in in vitro incubated human skeletal muscle. Diabetologia 34:899–901

Chapter 10
Spontaneously Immortalized Adult Rodent Schwann Cells as Valuable Tools for the Study of Peripheral Nerve Degeneration and Regeneration

Kazunori Sango, Masami Tsukamoto, Kazunori Utsunomiya, and Kazuhiko Watabe

Abstract We have established spontaneously immortalized Schwann cell lines from normal adult mice and rats, as well as murine disease models. One of the normal mouse cell lines, IMS32, possesses some biological properties of mature Schwann cells and high proliferative activities. The IMS32 cells have been utilized to investigate the action mechanisms of various molecules involved in peripheral nerve regeneration [e.g., ciliary neurotrophic factor (CNTF), sonic hedgehog, and galectin-1], and the pathogenesis of diabetic neuropathy, particularly the polyol pathway hyperactivity. The cell lines derived from murine disease models (e.g., lysosomal storage diseases, Charcot-Marie-Tooth disease, and neurofibromatosis) retain genomic and biochemical abnormalities, sufficiently representing the pathological features of the mutant mice. A normal rat cell line, IFRS1, retains the characteristic features of mature Schwann cells and the fundamental ability to myelinate axons in coculture with adult rat DRG neurons and PC12 cells. These Schwann cell lines can be valuable tools for exploring neuron–Schwann cell interactions, the

K. Sango (✉) • K. Watabe
ALS/Neuropathy Project, Department of Sensory and Motor Systems, Tokyo Metropolitan Institute of Medical Science, 2-1-6 Kamikitazawa, Setagaya-ku, Tokyo 156-8506, Japan
e-mail: sango-kz@igakuken.or.jp

M. Tsukamoto
Laboratory of Peripheral Nerve Pathophysiology (ALS/Neuropathy Project), Department of Sensory and Motor Systems, Tokyo Metropolitan Institute of Medical Science, 2-1-6 Kamikitazawa, Setagaya-ku, Tokyo 156-8506, Japan

Division of Diabetes, Metabolism and Endocrinology, Department of Internal Medicine, Jikei University School of Medicine, Minato-ku, Tokyo 105-8461, Japan

K. Utsunomiya
Division of Diabetes, Metabolism and Endocrinology, Department of Internal Medicine, Jikei University School of Medicine, Minato-ku, Tokyo 105-8461, Japan

K. Sango and J. Yamauchi (eds.), *Schwann Cell Development and Pathology*,
DOI 10.1007/978-4-431-54764-8_10, © Springer Japan 2014

pathobiology of axonal degeneration and regeneration in the peripheral nervous system, and novel therapeutic approaches against neurological disorders in patients with relevant diseases.

Keywords Adult rodents • Axonal regeneration • Immortalized Schwann cells • Murine disease models • Myelination • Peripheral neuropathies

10.1 Introduction

Schwann cells are essential components of the peripheral nervous system (PNS) in both vertebrates and invertebrates and play a multifunctional role in health, disease, and repair following nerve injury (Bunge 1993; Magnaghi et al. 2009). During their development and regeneration, Schwann cells proliferate, migrate, produce various factors that support the growth and maintenance of neurons, and ensheath their axons in either a myelinating or an unmyelinating form (Mirsky and Jessen 2007). The peripheral myelin, the ultimate structure of differentiation from myelinating Schwann cells in concert with their ensheathing axons, is indispensible to saltatory conduction. In addition, Schwann cells are suggested to play a key role in the regulation of immune responses under normal and pathological conditions in the PNS (Kieseier et al. 2007).

Cultured Schwann cells can be valuable tools for exploring the fundamental properties of Schwann cells in various stages of life from development to aging, the repair process following axonal injury, the pathogenesis of neurodegenerative disorders, and strategies for its prevention and treatment. Because some biological features of Schwann cells change with maturation and aging (Chi et al. 1993; Verdú et al. 2000), culture systems of mature Schwann cells appear to mimic axonal degeneration and regeneration better than those of immature cells. However, compared with a considerable number of articles on Schwann cells from embryonic and neonatal animals (Ogata et al. 2004; Yamauchi et al. 2008), surprisingly fewer studies have been conducted with primary cultures of Schwann cells from adult animals (Li 1998; Watabe et al. 1994; Haastert et al. 2007). One of the reasons for this appears to be the difficulty in isolating Schwann cells from mature peripheral nerves. Fully developed epineurium and perineurium with substantial amounts of connective tissue make it difficult and time consuming to obtain good yields of Schwann cells and sufficiently eliminate fibroblasts from the culture. To avoid such a long process of primary culture, cell lines have been established from schwannoma cells (Mizisin et al. 1996) and long-term Schwann cell cultures via transfection of oncogenes such as SV40 large T-antigen (Peden et al. 1989; Watabe et al. 1990) or spontaneous immortalization (Porter et al. 1987; Eccleston et al. 1991; Bolin et al. 1992; Toda et al. 1994; Watabe et al. 1995; Sango et al. 2011a). These lined cells rapidly proliferate and are more suitable for molecular and biochemical analyses than primary cultured Schwann cells. In contrast, the degree of differentiation and phenotypic expression of these Schwann cell lines differ from each other;

continuous cell lines that possess distinct phenotypes of mature Schwann cells are desirable (De Vries and Boullerne 2010). In this regard, Schwann cell lines established from adult animals via spontaneous immortalization appear to have advantages over those established via gene transduction. In this chapter, we summarize the characteristic features of adult rodent Schwann cell lines that we have established and discuss their usefulness for studying peripheral nerve degeneration and regeneration.

10.2 How to Establish Spontaneously Immortalized Schwann Cells?

10.2.1 Biological Basis for Schwann Cell Immortalization

Although the detailed mechanism of spontaneous Schwann cell immortalization remains unknown, rat Schwann cells can reportedly divide indefinitely under appropriate culture conditions (Mathon et al. 2001). Up to now, we have been able to obtain spontaneously immortalized Schwann cell lines from ICR (Watabe et al. 1995), BALB/c (Watabe et al. 2003), and C57BL mouse strains (Watabe et al. 2001; Shen et al. 2002; Ohsawa et al. 2005; Kawashima et al. 2007) and Fischer 344 (Sango et al. 2011a), Wistar, and Sprague–Dawley rat strains (Watabe et al., unpublished data) (Table 10.1). Considering these findings, it is likely that the spontaneous immortalization of long-term cultured Schwann cells is a general phenomenon in rodents, irrespective of strain. Because both short-term cultured and immortalized

Table 10.1 Spontaneously immortalized adult rodent Schwann cell lines established at the authors' institutions

Line	Origin (disease model)	Strain	References
IMS32	Wild-type mice	ICR	Watabe et al. (1995)
SPMS9	*spm/spm* (Niemann–Pick disease type C)	C57BL/KsJ	Watabe et al. (2001)
573C10	npc^{nih}/npc^{nih} (Niemann–Pick disease type C)	BALB/c	Watabe et al. (2003)
574C3	$npc^{nih}/+$	BALB/c	Watabe et al. (2003)
TwS1	Twitcher (globoid cell leukodystrophy (Krabbe))	C57/BL6J	Shen et al. (2002)
1113C1	*Hexb−/−* (G_{M2} gangliosidosis (Sandhoff))	C57/BL6	Ohsawa et al. (2005)
1089C1	*α-Gal A* (*−/0*) (Fabry disease)	C57/BL6	Kawashima et al. (2007)
675C20	*P0−/−* (Charcot-Marie-Tooth disease type 1B)	C57/BL6	Watabe et al. (2003)
676C2	*P0+/−*	C57/BL6	Watabe et al. (2003)
677C1	*P0+/+* (wild-type mice)	C57/BL6	Watabe et al. (2003)
654C1	$Nf1^{Fcr}/c$ (neurofibromatosis type I)	C57/BL6	Watabe et al. (2003)
IFRS1	Wild-type rats	Fischer344	Sango et al. (2011a, b)

Schwann cells secrete autocrine growth factors [e.g., platelet-derived growth factor (PDGF), transforming growth factor-β (TGF-β)], Schwann cells with an enhanced capacity to respond to growth factors may be selected through repetitive passaging (Porter et al. 1987). In other studies (Muir et al. 1990; Eccleston et al. 1991), antiproliferative activity was detected in conditioned medium (CM) from short-term cultured rat Schwann cells but not from long-term cultured or immortalized Schwann cells. These findings suggest that Schwann cell immortalization is attributable, at least partially, to the loss of an autocrine growth inhibitory loop.

The key to establishing spontaneously immortalized Schwann cells is to eliminate neurons, fibroblasts, and other cells from the mixed primary culture. We successfully obtained Schwann cell lines from the primary cultures of dorsal root ganglia (DRG) and adjacent peripheral nerves from adult mice and rats. Neurons lack proliferative activity and gradually disappear from the culture, whereas fibroblasts are a major contaminant of these preparations. Antiserum against the fibroblast surface antigens Thy-1.1 (rats) and Thy-1.2 (mice) and complement have been employed for the removal of fibroblasts from the primary culture (Brockes et al. 1981). In another approach to purify Schwann cells, the primary culture is maintained in serum-free culture conditions with the use of Schwann cell mitogens, such as neuregulin (NRG)-β and forskolin (Haastert et al. 2007). In addition to the elimination of serum that effectively suppresses the proliferation of fibroblasts, treatment with NRG-β promotes continuous growth and subculture of both mouse and rat Schwann cells. Although forskolin activates adenylate cyclase and elevates intracellular cyclic AMP levels in both species, it suppresses the mitogenic response of mouse Schwann cells to growth factors [e.g., PDGF, fibroblast growth factor (FGF), TGF-β], in sharp contrast to its promoting activity on the proliferation of rat Schwann cells (Watabe et al. 1994, 1995; Sango et al. 2011a). Why the mitogenic activity of forskolin on cultured Schwann cells differs between these species remains unknown.

The following sections provide a brief description of the establishment of two representative Schwann cell lines: *IMS32* from adult ICR mice and *IFRS1* from adult Fischer 344 rats. The biological properties of these cell lines is discussed in Sects. 10.3 and 10.5 in this chapter.

10.2.2 Establishment of IMS32 Cells: Serum-Containing Culture with Fibroblast Elimination Using Anti-Thy-1.2 and Complement

Primary and long-term cultures of Schwann cells were prepared from DRG and adjacent peripheral nerves derived from adult ICR mice and maintained in serum-containing medium. Following repeated treatment of the primary cultures with antibody against mouse Thy-1.2 and rabbit complement for the first 2 to 3 weeks, most fibroblasts were eliminated, and the cultures consisted of more than 95 % Schwann cells.

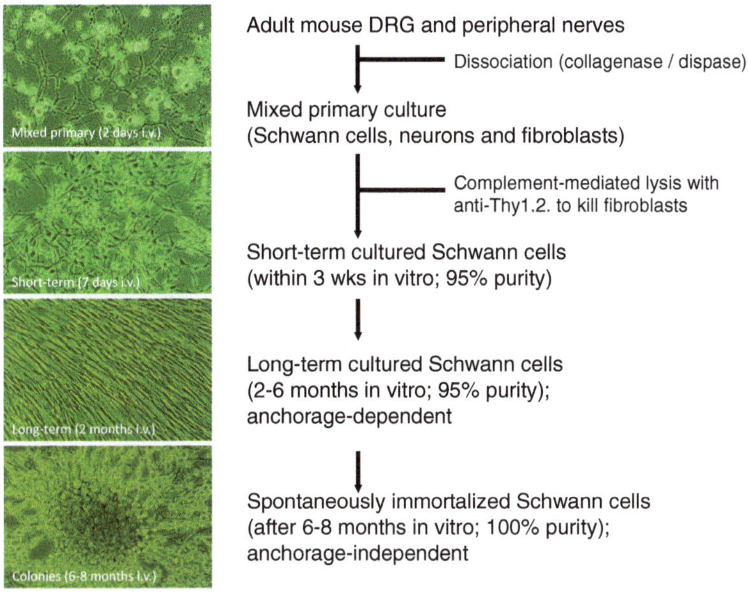

Fig. 10.1 Establishment of spontaneously immortalized adult mouse Schwann cells: serum-containing culture with the elimination of fibroblasts by anti-Thy1.2 and complements. *DRG* dorsal root ganglia. (Phase-contrast micrographs reproduced from Watabe et al. (2003) *Neuropathology* 23(1):68–78, with permission from John Wiley & Sons publications)

These cells were fed twice a week and passaged once every 2 weeks. After 6 to 8 months in vitro, spontaneously developed colonies were observed (Fig. 10.1). They were separated using cloning rings, and five different cell lines (IMS8, -13, -26, -29, and -32) were obtained. One of the cell lines, IMS32, was further characterized (Watabe et al. 1995).

10.2.3 Establishment of IFRS1 Cells: Serum-Free Culture Supplemented with NRG-β and Forskolin

Primary and long-term cultures of Schwann cells from DRG and peripheral nerves of adult Fischer 344 rats were grown in serum-free medium containing 40 ng/ml NRG-β and 5 μM forskolin, which was changed twice a week, and were passaged once every 4 to 6 weeks, during which time neurons and fibroblasts ceased to grow in the cultures. After 4 to 5 months in culture, spontaneously emerging Schwann cell colonies were isolated using cloning rings and further expanded (Fig. 10.2). One of these cell lines, designated as IFRS1, was further characterized (Sango et al. 2011a).

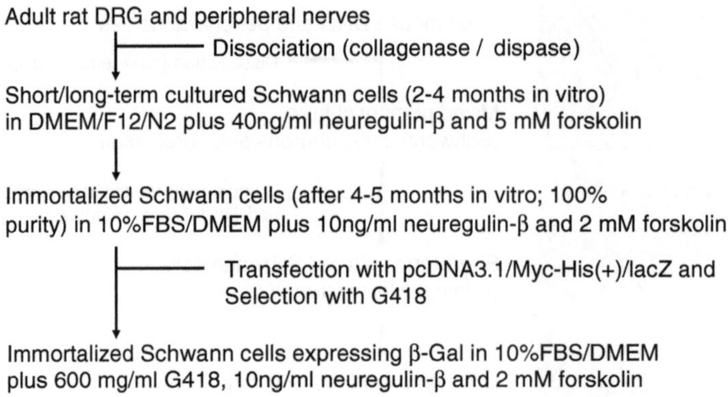

Fig. 10.2 Establishment of spontaneously immortalized adult rat Schwann cells in serum-free culture supplemented with neuregulin-β and forskolin

10.3 Immortalized IMS32 Adult Mouse Schwann Cells

10.3.1 Biological Features of IMS32 Cells

IMS32 cells appear to be one of the best characterized Schwann cell lines at present. They display distinct Schwann cell phenotypes such as spindle-shaped morphology and the expression of glial cell markers [e.g., S100, glial fibrillary acidic protein (GFAP), p75 low-affinity neurotrophin receptor ($p75^{NTR}$)] (Fig. 10.3), transcription factors [e.g., PAX3, Krox20, Oct6, Sox10], and myelin proteins [e.g., P0, peripheral myelin protein 22 kDa (PMP22)] crucial for Schwann cell development and peripheral myelin formation, and neurotrophic factors required for the survival of neurons and maintenance of neuron–Schwann cell interactions [e.g., nerve growth factor (NGF), brain-derived neurotrophic factor (BDNF), neurotrophin-3, glial cell line-derived neurotrophic factor (GDNF), ciliary neurotrophic factor (CNTF)] (Watabe et al. 1995, 2003). Similar to primary and long-term cultured Schwann cells, IMS32 cells exhibit mitogenic responses to several growth factors [e.g., PDGF-BB, acidic and basic FGF, TGF-β1, TGF-β2, NRG-β].

One of the fundamental properties of Schwann cells is the production and secretion of various neurotrophic factors and cytokines that promote neuronal survival and axonal regeneration following axonal injury. We observed that CM obtained from IMS32 cells enhanced the neurite elongation of PC12 rat pheochromocytoma cells. The neurite outgrowth-promoting activity of CM was not attenuated by cotreatment with anti-NGF neutralizing antibody, suggesting the combined effects of multiple neurotrophic factors, other than NGF, in the case of CM. Western blot analysis revealed the presence of several neurotrophic factors and cytokines such as NGF, TGF-β, and galectin-1 (GAL-1) in the CM (Watabe et al. 1995; Sango et al. 2004).

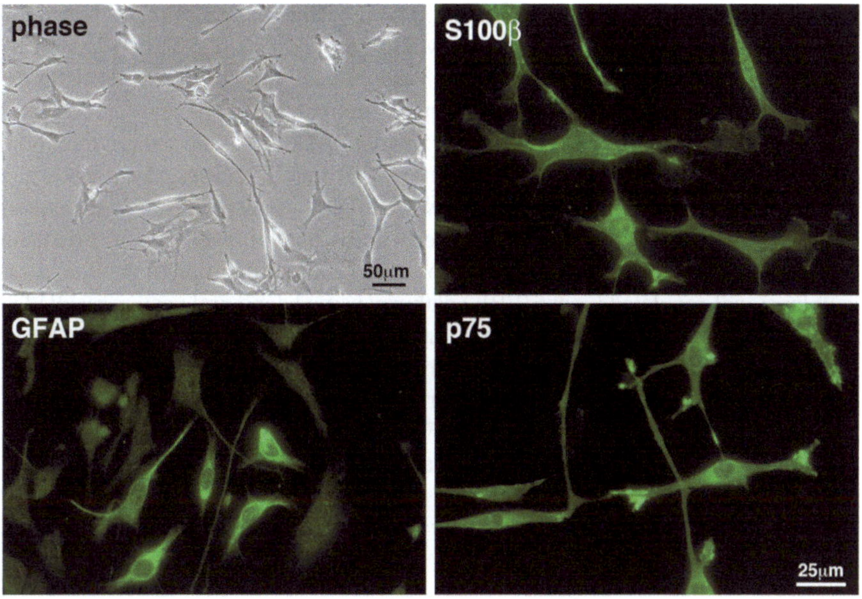

Fig. 10.3 IMS32 cells displayed distinct Schwann cell phenotypes such as spindle-shaped morphology under a phase-contrast microscope and immunoreactivity to *S100β*, *GFAP*, and the *p75* low-affinity neurotrophin receptor

In contrast to these Schwann cell-like phenotypes, IMS32 cells differ from primary and long-term cultured Schwann cells in that the former were not contact inhibited and that they formed ball-shaped subcolonies when cultures reached confluence (Watabe et al. 1995; Sango et al. 2004). We failed to demonstrate that the cell line could myelinate a mouse axon in the same manner as endogenous Schwann cells in the peripheral nerves and primary cultured Schwann cells. The high proliferative activity of IMS32 cells may impede continuous and stable neuron–Schwann cell interactions, which usually take 4 weeks or longer to form the myelin sheath. Despite these differences from normal Schwann cells, IMS32 cells have been utilized to investigate the action mechanisms of various molecules involved in peripheral nerve regeneration and the pathogenesis of peripheral neuropathy.

10.3.2 IMS32 for the Study of Axonal Regeneration-Related Molecules

10.3.2.1 CNTF

CNTF is abundantly expressed in peripheral nerves with immunoreactivity localized to the myelinating Schwann cells of adult rodents, and it is recognized as a trophic molecule for a wide variety of neurons (Adler et al. 1979; Sendtner et al. 1992;

Sango et al. 2008a). Ito et al. (2006) observed that suppression of the transcription factor Sox10 by RNA interference in IMS32 cells reduced CNTF expression by more than 80 %. This finding agrees with the reduced CNTF protein levels in the sciatic nerves of mice with heterozygous inactivation of the Sox10 gene and indicates that Sox10 is necessary and sufficient for regulating CNTF expression in the PNS.

Despite its potent neuroprotective activity in injured neurons, the mRNA and protein levels of CNTF are markedly downregulated in nerves distal to the site of injury, and its expression does not recover without axonal regeneration (Friedman et al. 1992). According to the study by Abe et al. (2001), inhibition of the Ras extracellular signal-regulated kinase (ERK) pathway using adenoviral vectors resulted in a significant increase in CNTF expression in IMS32 cells. They also observed that increased ERK phosphorylation was correlated with the downregulation of CNTF expression in the axotomized sciatic nerve. Taking these findings together, ERK inactivity appears to be crucial for CNTF expression in Schwann cells, and activation of ERK following nerve injury critically influences the expression of CNTF.

10.3.2.2 Sonic Hedgehog (Shh)

Shh belongs to the hedgehog family and plays a key role in embryonic tissue induction and patterning. Reportedly, Shh is a therapeutic target for various neurodegenerative disorders, including diabetic neuropathy (Calcutt et al. 2003; Traiffort et al. 2010); however, little is known about its functional role in axonal regeneration following injury. Hashimoto et al. (2008) observed upregulation of Shh expression in the Schwann cells of crush-injured adult rat sciatic nerves. In addition, treatment of primary cultured Schwann cells with exogenous Shh enhanced the mRNA expression of BDNF. To further investigate whether Shh regulates BDNF promoter activities, they performed luciferase reporter assays using IMS32 cells because primary cultured Schwann cells are resistant to lipofection. BDNF promoter activity was increased in IMS32 cells cotransfected with Shh and its receptor Smoothened. These findings imply that Shh plays a neuroprotective role, at least in part, through BDNF.

10.3.2.3 GAL-1

GAL-1, the first protein identified in a family of β-galactoside-binding animal lectins, is suggested to play a role in the development and regeneration of the nervous system (Camby et al. 2006; Sango et al. 2012a). Despite lacking a signal leading peptide, GAL-1 is subject to externalization via a nonclassical pathway from various types of cells (Cooper and Barondes 1990). Our immunocytochemical and Western blot analyses revealed the externalization of GAL-1 from primary cultured adult rat DRG neurons and Schwann cells as well as IMS32 cells (Sango et al. 2004). Following externalization, some GAL-1 molecules associate with surface or extracellular matrix glycoconjugates, in which lectin activity is stabilized, whereas free

GAL-1 molecules are rapidly oxidized in the nonreducing extracellular environment. Growing evidence suggests that both the reduced and oxidized forms of GAL-1 are involved in the repair process after nerve injury (Camby et al. 2006); however, there is a marked difference in the structural and functional properties between the two forms; GAL-1 in the reduced form acts on nervous tissue as a lectin (Sasaki et al. 2004; Plachta et al. 2007), whereas GAL-1/Ox lacks lectin activity but could promote axonal regeneration and Schwann cell migration via activating macrophages (Horie et al. 1999, 2004; Sango et al. 2012a).

10.3.3 IMS32 Cells as a Valuable Tool for Studying Peripheral Neuropathy

10.3.3.1 Diabetic Neuropathy

As described in Chap. 9, Schwann cell abnormalities caused by hyperglycemia or hyperlipidemia can be a cause of nerve dysfunction such as reduced nerve conduction velocity, axonal atrophy, and impaired axonal regeneration (Dyck and Giannini 1996; Eckersley 2002). Although the detailed pathogenesis remains unclear, metabolic alterations in Schwann cells under hyperglycemic or hyperlipidemic conditions appear to be, at least in part, associated with polyol pathway hyperactivity, glycation of cellular proteins and lipids, increased oxidative stress, altered protein kinase C activity, and a reduced supply of neurotrophic factors. Moreover, impaired lipid and cholesterol metabolism in Schwann cells under diabetic conditions may affect the structure and function of peripheral myelins (Lehmann and Höke 2010). Because we reported polyol pathway hyperactivity in IMS32 under high-glucose conditions (Sango et al. 2006b), this cell line has been repeatedly employed for studying diabetic neuropathy (Ota et al. 2007; Tosaki et al. 2008; Sango et al. 2008b; Suzuki et al. 2011). The major findings from those reports have been fully discussed elsewhere (Sango et al. 2011b); thus, we only mention the essential points here.

Polyol Pathway

Aldose reductase (AR), the first enzyme in the polyol pathway, is localized to Schwann cells in the peripheral nerves, and the increased glucose flux into the pathway via AR and the subsequent accumulation of sorbitol in Schwann cells can directly or indirectly affect nerve functions (Song et al. 2003). We observed increased AR mRNA/protein expression and marked sorbitol and fructose accumulation in IMS32 cells cultured under a high-glucose (30 mM) condition. Furthermore, application of an AR inhibitor, fidarestat (Sanwa Kagaku Kenkyusho, Nagoya, Japan), to the high-glucose medium diminished the intracellular sorbitol content to a level close to that in a normal glucose (5.6 mM) medium (Sango et al. 2006b).

These findings suggest that the culture of IMS32 cells under high-glucose conditions is a suitable in vitro model for studying polyol pathway-related abnormalities in diabetes.

Glycation

Advanced glycation end product (AGE)-induced modification of myelin proteins, cytoskeletal proteins, and extracellular matrix proteins contributes to segmental demyelination, axonal degeneration, and impaired axoplasmic transport and regenerative activity (Sugimoto et al. 2008). Methylglyoxal (MG) is a reactive dicarbonyl precursor of AGEs and a potent source of reactive oxygen species (ROS). Exogenously applied MG exhibited potent cytotoxicity in primary cultured Schwann cells (Fukunaga et al. 2004) and IMS32 cells (Ota et al. 2007; Sango et al. 2008b). Metformin, an oral hypoglycemic biguanide, inhibited the MG-induced activation of caspase-3 and c-Jun-N-terminal kinase (JNK), intracellular ROS formation, and apoptosis in IMS32 cells (Ota et al. 2007). These findings suggest that metformin potentially prevents diabetic neuropathy and other complications by reducing the formation of AGEs and ROS, in addition to lowering the blood glucose level.

Oxidative Stress

We observed that high-glucose (30 and 56 mM) conditions induced the upregulation of oxidative stress markers such as 4-hydroxy-2-nonenal, acrolein, and hexanoyl lysine, but not apoptotic cell death. in IMS32 cells (Sango et al. 2008b). Although oxidative damage caused by mitochondrial dysfunction under hyperglycemic conditions may trigger the apoptotic cascade (Nishikawa and Araki 2007), the load of high glucose on Schwann cells does not appear to be a sufficient inducer of cell death in the absence of further insults. For instance, palmitic acid-induced lipotoxicity accelerated the apoptotic cascade in primary cultured and immortalized Schwann cells under normal and high-glucose conditions (Padilla et al. 2011; Suzuki et al. 2011).

Reduced Synthesis of Neurotrophic Factors

A reduced supply of NGF to the sensory and sympathetic neurons can be a cause of small sensory and autonomic fiber dysfunction and impaired axonal regeneration after injury in diabetes (Sango et al. 1994, 2006a). Tosaki et al. (2008) reported that adult mouse DRG neurons cultured in CM from IMS32 cells under high-glucose (30 mM) conditions exhibited lower NGF concentrations and neurite outgrowth activity than those under normal glucose (5.5 mM) conditions. Therefore, reduced NGF synthesis by Schwann cells under hyperglycemic conditions can be a cause of impaired axonal regeneration and dysfunction of small-fiber sensory and autonomic fibers.

10.3.3.2 Amyloid Polyneuropathy

As described in Chap. 7, familial amyloid polyneuropathy (FAP) is an autosomal dominant neurodegenerative disorder characterized by the systemic deposition of mutant transthyretin (TTR) amyloid fibrils, particularly in the PNS (Sousa and Saraiva 2003). TTR mRNA is present in the liver, choroid plexus of the brain, retinal pigmented epithelium of the eye, pancreatic islets, and meninges, although it is unclear whether TTR mRNA is expressed in the PNS. Murakami et al. (2010) demonstrated by reverse transcription-polymerase chain reaction (RT-PCR) that TTR mRNA can be detected in mouse sciatic nerves, primary cultured Schwann cells, and IMS32 cells. These findings suggest a role for TTR in axonal regeneration (Fleming et al. 2009). They also observed human TTR mRNA/protein in cultured transgenic mouse Schwann cells, suggesting that the sources of TTR amyloid deposits in the PNS include Schwann cells in addition to the liver and choroid plexus.

10.4 Immortalized Schwann Cells from Murine Disease Models

The establishment of Schwann cell lines from murine disease models may greatly facilitate research of the cellular mechanisms of their PNS lesions in relevant diseases. In addition to spontaneously immortalized Schwann cell lines derived from normal (wild-type) mice, such as IMS32 cells, we have established Schwann cell lines from murine models of Niemann–Pick disease type C (NPC) (Watabe et al. 2001, 2003), Krabbe disease (Shen et al. 2002), Sandhoff disease (Ohsawa et al. 2005), Fabry disease (Kawashima et al. 2007), Charcot-Marie-Tooth disease type 1B (CMT1B), and neurofibromatosis (Watabe et al. 2003) (Table 10.1). These cell lines retain genomic and biochemical abnormalities, sufficiently representing the pathological features of the mutant mice.

10.4.1 Lysosomal Storage Diseases

Lysosomal storage disorders are defined by the defective activity of lysosomal enzymes, which catalyze the degradation of metabolites. The diseases are categorized according to the nature of the accumulated metabolites, such as sphingolipidosis, mucopolysaccaridosis, and glycoprotein storage diseases. Most genes encoding lysosomal enzymes have been cloned and the disease-causing mutations have been characterized. Although few naturally occurring mouse models of human lysosomal diseases are known, numerous models of lysosomal diseases have been generated by targeted gene disruption (Suzuki et al. 1999).

10.4.1.1 NPC

NPC is an autosomal recessive disorder characterized by the defect of intracellular transport of cholesterol and the accumulation of unesterified cholesterol in the endosomal/lysosomal system, causing progressive neurodegeneration and death during early childhood. The genes responsible for the disease, *NPC1* and *HE1*, have been identified (Carstea et al. 1997; Naureckiene et al. 2000). Two mouse models of NPC, BALB/c npc^{nih}/npc^{nih} (Pentchev et al. 1980) and C57BL/KsJ sphingomyelinosis (*spm/spm*) (Miyawaki et al. 1982), displayed markedly reduced *Npc1* mRNA expression (Loftus et al. 1997) with the development of hepatosplenomegaly and ataxic movements at the age of 7 to 8 weeks. The Schwann cell lines established from these mice, designated 573C10 and SPMS9, contain intracytoplasmic granules that are positive for filipin cholesterol staining and immunoreactive for G_{M2} ganglioside (Watabe et al. 2001, 2003). Treatment with an inhibitor of ceramide-specific glucosyltransferase, *N*-butyldeoxynojirimycin, markedly reduced the intracytoplasmic granular immunofluorescence for G_{M2} ganglioside in SPMS9 cells.

10.4.1.2 Krabbe Disease (Twitcher)

The twitcher mouse is a murine model of human globoid cell leukodystrophy (GLD; Krabbe disease) caused by a genetic defect in the activity of galactosylceramidase (GALC) (Kobayashi et al. 1980). The accumulation of a cytotoxic metabolite, galactosylsphingosine (psychosine), in myelin-forming cells (oligodendrocytes and Schwann cells) of the twitcher mouse as well as patients with GLD has been suggested to cause dysfunction of these cells and subsequent demyelination in the central nervous system and PNS. One of the Schwann cell lines established from twitcher mice, designated TwS1, exhibited markedly reduced GALC activity and elevated psychosine levels (Shen et al. 2002). Ultrastructurally, varieties of cytoplasmic inclusions were demonstrated in TwS1 cells. When TwS1 cells were infected with a retrovirus vector encoding GALC, GALC activity was markedly increased and psychosine levels were significantly decreased.

10.4.1.3 Sandhoff Disease

Tay–Sachs and Sandhoff diseases are autosomal recessive neurodegenerative disorders characterized by deficient lysosomal β-hexosaminidase activity, resulting in the progressive accumulation of G_{M2} ganglioside and related substrates in the nervous system (Sandhoff 2001). A mouse model of Sandhoff disease created via disruption of the mouse *Hexb* gene displays progressive neurological manifestations similar to patients with the disease (Sango et al. 1995). One of the Schwann cell lines established from Sandhoff mice, designated 1113C1, displayed increases in the size and number of lysosomes with massive G_{M2} ganglioside accumulation (Ohsawa et al. 2005). Incorporation of recombinant human β-hexosaminidase isozymes expressed in Chinese hamster ovary cells into the cultured Sandhoff Schwann cells via cation-independent mannose-6-phosphate receptors resulted in the degradation of the accumulated G_{M2} ganglioside.

10.4.1.4 Fabry Disease

Fabry disease is an X-linked recessive disorder characterized by a deficiency of α-galactosidase A, which cleaves the terminal molecule of galactose from the accumulating ceramide trihexoside. Neutral glycosphingolipids with terminal α-linked galactosyl moieties, predominantly globotriaosylceramide, accumulate in various organs and tissues of patients with Fabry disease; the storage of these moieties in the PNS results in severe pain in the peripheral extremities, acroparesthesia, and disorders of the autonomic nervous system (Banerjee 2004). One of the Schwann cell lines established from the murine model of Fabry disease (Ohshima et al. 1997), designated 1089C1, exhibited deficient α-galactosidase activity and numerous cytoplasmic inclusion bodies (Kawashima et al. 2007). Recombinant α-galactosidase added to the culture medium was incorporated into 1089C1 cells in a concentration-dependent manner; however, the increase in cell-associated α-galactosidase activity was less than that in mouse and human Fabry fibroblasts. The lower enzyme uptake can be attributed to the low expression of cation-independent mannose-6-phosphate receptors in Schwann cells.

10.4.2 CMT1B

As described in Chap. 6, Charcot-Marie-Tooth disease is probably the most common inherited neurological disorder. CMT1B is caused by mutations in the major myelin protein zero (MPZ or P0), which accounts for more than half the total PNS myelin protein. The P0 protein is exclusively expressed by myelinating Schwann cells as the most prominent component of compact myelin, and this protein has been suggested to play a major role in the formation and maintenance of the myelin sheath (Arroyo and Scherer 2007). P0-deficient mice [*P0(−/−)*] generated by targeted gene disruption are severely hypomyelinated with predominantly uncompacted myelin, are deficient in normal motor coordination, and exhibit tremors and occasional convulsions (Giese et al. 1992). We have established Schwann cell lines from *P0(−/−)*, *P0(+/+)*, and *P0(+/−)* (Watabe et al. 2003). These cell lines can be useful tools for investigating the detailed pathogenesis of CMT1B and the mechanisms of complex myelin gene regulation in vitro.

10.4.3 Neurofibromatosis Type I (NF1)

NF1, or von Recklinghausen neurofibromatosis, which is one of the most common dominant genetic disorders, is characterized by the development of multiple benign and malignant nervous system tumors such as neurofibroma, malignant peripheral nerve sheath tumor (MPNST), and glioma (Cichowski and Jacks 2001). The NF1 gene, which is mutated in NF1 disease, encodes neurofibromin, which harbors a functional Ras-GTPase-activating protein (GAP) homologous domain and negatively regulates the activation of the Ras intracellular signaling pathway. We established Schwann cell lines from heterozygous NF1-deficient mice ($Nf1^{Fcr}/+$) but not from homozygous mice ($Nf1^{Fcr}/\ Nf1^{Fcr}$) because $Nf1^{Fcr}/Nf1^{Fcr}$ mice die in utero of cardiac abnormalities (Brannan et al. 1994). The 654C1 cells heterozygous for the NF1 gene exhibited markedly reduced neurofibromin expression (Watabe et al. 2003) and decreased Ras-GAP activity (Araki et al., unpublished observation).

10.5 Immortalized IFRS1 Adult Rat Schwann Cells

10.5.1 Biological Features of IFRS1 Cells

IFRS1 cells displayed distinct Schwann cell phenotypes such as spindle-shaped morphology and intense immunoreactivity for S100β, GFAP, and vimentin (Fig. 10.4). IFRS1 cells express transcription factors (Krox20, Oct6, SOX10) and myelin proteins (P0, PMP22, MAG) crucial for Schwann cell development and peripheral myelin formation, in addition to neurotrophic factors (NGF, GDNF, CNTF), neurotrophin receptors (truncated TrkB, TrkC), and cell adhesion molecules (L1, NCAM, N-cadherin) required for the survival and neurite outgrowth of neurons and maintenance of neuron–Schwann cell interactions. CM obtained from IFRS1 cells promoted the viability and neurite outgrowth of adult rat DRG neurons, suggesting that the cells produce and secrete neurotrophic molecules required for neuronal survival and axonal regeneration after injury.

10.5.2 Myelination in Co-culture of IFRS1 Cells with Neurons

Peripheral myelin is formed via complicated interplay between neurons and Schwann cells during PNS development and regeneration. Co-cultures of neurons and Schwann cells have been used as models to investigate the mechanisms of myelination under normal and pathological conditions. There are many reports on myelination in co-culture of primary cultured neurons and primary cultured Schwann cells (Kleitman et al. 1998; Gingras et al. 2008; Lehmann et al. 2009; Zhang et al. 2009). Although these models are beneficial for studying neuron–Schwann cell interactions, the process requires killing the animals for every primary

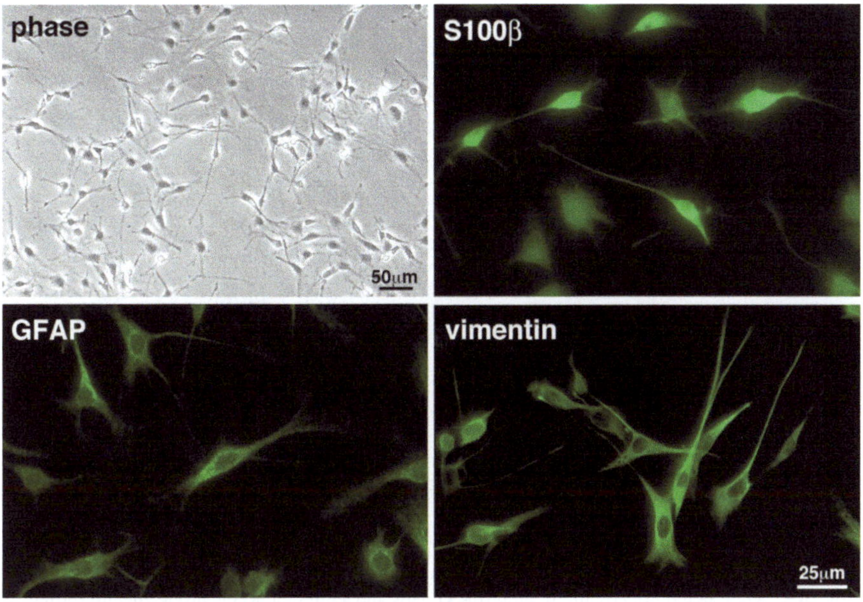

Fig. 10.4 IFRS1 cells exhibited distinct Schwann cell phenotypes such as spindle-shaped morphology under a phase-contrast microscope and immunoreactivity to S100β, GFAP, and vimentin

culture preparation. Therefore, it would be desirable to establish stable co-culture systems with neuronal or Schwann cell lines. In our initial attempt, co-cultures of IMS32 cells with PC12 cells or adult mouse DRG neurons failed to exhibit myelination (Watabe et al., unpublished observation). The high proliferative activity of IMS32 cells may impede continuous and stable neuron–Schwann cell interactions, which usually take 4 weeks or longer to form the myelin sheath. In contrast to the rapid proliferation of IMS32 cells even in the absence of exogenous growth stimulants, NRG-β and forskolin are required for the growth and passage of IFRS1 cells. In other words, overgrowth of IFRS1 cells can be prevented by the absence of these stimulants in coculture with neurons. We succeeded in establishing stable co-culture systems with IFRS1 cells and primary cultured DRG neurons (Sango et al. 2011a) or NGF-primed PC12 cells (Sango et al. 2012b) and confirmed the myelin formation by light and electron microscopy. These findings suggest that IFRS1 cells retain the fundamental ability to myelinate axons, thereby being a valuable tool for exploring neuron–Schwann cell interactions.

10.5.2.1 Adult Rat DRG Neurons

IFRS1 cells were co-cultured with adult rat DRG neurons in serum-free medium (F12/B27) supplemented with 50 μg/ml ascorbic acid and 10 ng/ml GDNF. After 28 days of coculture, myelin formation was illustrated by double immunofluorescence

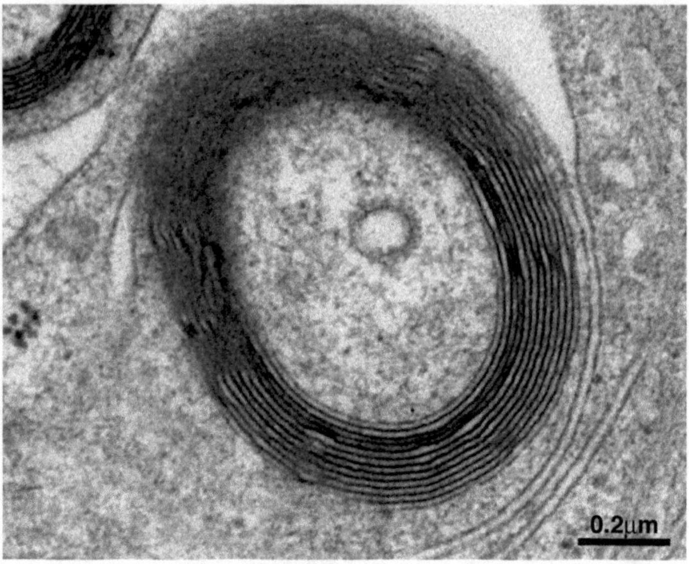

Fig. 10.5 After 28 days of co-culture with IFRS1 cells and adult rat DRG neurons, myelin formation is illustrated by electron microscopy

staining with antibodies against P0 and βIII tubulin, Sudan black B staining, and electron microscopy (Fig. 10.5). Compared with the coculture models used in previous studies, the DRG neuron–IFRS1 co-culture system has the following advantages:

1. Both neurons and Schwann cells are derived from adult animals, and they possess biological properties of the mature PNS
2. Spontaneously immortalized Schwann cells can be stably and effectively used in co-culture

Although the DRG neuron–IFRS1 co-culture system has advantages, animals must be killed for every preparation of the primary culture of DRG neurons. The primary culture contains a small number of nonneuronal cells, including Schwann cells, and we cannot deny the possibility that primary cultured Schwann cells can form myelin in the co-culture system. Then, we inserted the β-galactosidase gene into the IFRS1 cells. These LacZ-labeled IFRS1 cells had the same Schwann cell phenotype as the original IFRS1 cells, and they were distinguished from the primary cultured Schwann cells by X-Gal staining. The histochemical findings indicate that the LacZ-labeled IFRS1 cells are capable of myelinating neurites. However, if we can establish stable co-culture systems of IFRS1 cells with pure neuronal cell lines, we will avoid these issues arising from the primary culture.

10.5.2.2 PC12 Cells

The protocol used for the co-culture of DRG neurons–IFRS1 cells did not work for that of PC12-IFRS1 cells, largely because of the excess proliferation of PC12 cells. Replacement of B27 with N2 supplement, which contains only five nutrients and no antioxidants, partially prevented the proliferation of PC12 cells but resulted in massive IFRS1 cell death by 3 weeks of coculture. Our trial-and-error experience with cultures helped generate possible solutions to these problems, and we finally succeeded in myelination under the following protocol.

1. Before the co-culture, PC12 cells were seeded at low density ($3 \times 10^2/cm^2$) and maintained in serum-containing medium for 1 day and then in serum-free medium with N2 supplement, ascorbic acid (50 μg/ml), and NGF (50 ng/ml) for 7 days. Exposure to such an NGF-rich environment with minimum nutrients accelerated the differentiation and neurite extension but not the proliferation of PC12 cells.
2. When IFRS1 cells were added to NGF-primed PC12 cells, the cell density ratio of PC12 cells to IFRS1 cells was adjusted from 1:100 to 1:200. Incubation of the coculture in serum-containing medium for the initial 2 days was required for IFRS1 cells to attach to the substrate and change their morphology from round to spindle shaped.
3. The co-cultured cells were then maintained in serum-free medium with B27 supplement, ascorbic acid (50 μg/ml), NGF (10 ng/ml), and recombinant human neuregulin-1 isotype SMDF (25 ng/ml). After 21 days of co-culture in "myelination medium," PC12 cells aggregated to form small clusters of neuronal cell bodies with a network. These morphological changes were similar to those in the co-culture of DRG neurons and IFRS1 cells, and they indicate stable and effective neuron–Schwann cell interactions. At day 28 of co-culture, myelin formation was illustrated by light and electron microscopy (Fig. 10.6).

SMDF increased the relative cell density of IFRS1 and the protein expression of P0 and ErbB3, whereas it decreased the protein expression of $p75^{NTR}$ (Sango et al. 2012b). These findings suggest that SMDF is required for the long-term survival of IFRS1 cells and myelination in the coculture system.

The PC12–IFRS1 coculture model has significant advantages compared with previous models using primary cultured neurons or Schwann cells, as follows.

1. It can be prepared at the researchers' convenience without the time-consuming processes of the primary culture.
2. It consists of pure neuronal and Schwann cell lines and does not include other cells derived from primary cultures.
3. It is free of ethical problems that may arise from sacrificing animals.
4. It can be prepared and maintained by routine culture techniques without genetic manipulation of cells.

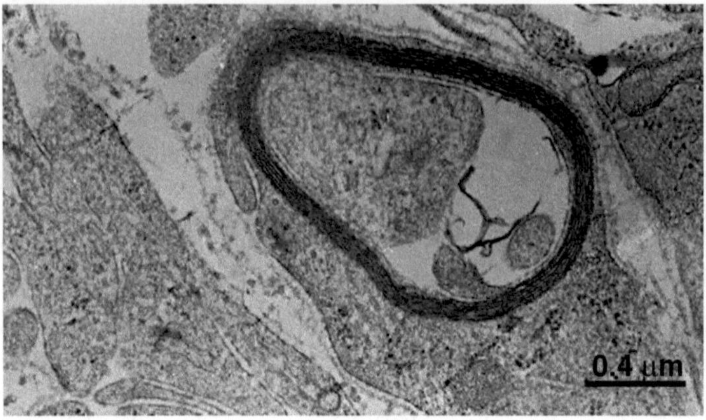

Fig. 10.6 After 28 days of co-culture with IFRS1 cells and NGF-primed PC12 cells, myelin formation is illustrated by electron microscopy

10.5.2.3 Motor Neurons

In addition to the DRG neuron–IFRS1 and PC12–IFRS1 co-culture systems, our current investigation is aimed at establishing co-culture models of IFRS1 cells and cell lines of motor neurons that would be more suitable than previous (Gingras et al. 2008) and present models to study motor neuron diseases such as amyotrophic lateral sclerosis, peripheral neuropathies such as Charcot-Marie-Tooth disease, multifocal motor neuropathies, and demyelinating diseases such as Guillain–Barré syndrome and chronic inflammatory demyelinating neuropathy.

10.5.3 IFRS1 Cells for the Study of Axonal Degeneration and Regeneration

Because IFRS1 cells were recently introduced (Sango et al. 2011a), only few studies have utilized these cells.

10.5.3.1 GAL-1

According to immunocytochemistry, GAL-1 was localized to the cell bodies and processes of IFRS1 cells. Furthermore, Western blot analysis revealed the intense immunoreactivity for GAL-1 in both IFRS1 cells and culture medium (Sango et al. 2012a). These findings suggest that IFRS1 cells synthesize and secrete GAL-1 in a similar manner to DRG neurons and IMS32 cells (Sango et al. 2004). In our recent study, recombinant GDNF promoted neurite outgrowth from adult rat DRG neurons and GAL-1 protein expression in the neuron-enriched culture of DRG. In contrast, GDNF failed to upregulate GAL-1 expression in IFRS1 cells (Takaku et al. 2013). The differential effects of GDNF may be caused by different GDNF signaling molecules and pathways between neurons and Schwann cells (Airaksinen and Saarma 2002; Paratcha et al. 2003). GDNF has been demonstrated to promote myelination in co-culture of DRG neurons and IFRS1 cells (Sango et al. 2011a); however, it failed to upregulate GAL-1 expression in the same coculture system (Takaku et al. 2013). Although further study is required, the GDNF/GAL-1 signaling axis appears to be more involved in initial axonal regeneration than in neuroprotection and remyelination during the repair process after peripheral nerve injury.

10.5.3.2 Diabetic Neuropathy

In contrast to IMS32 cells, neither AR expression nor intracellular polyol levels were enhanced by exposure of IFRS1 cells to a high-glucose (30 mM) condition (Tsukamoto and Sango, unpublished data). However, our preliminary study revealed that high-glucose conditions induced the upregulation of galectin-3 (GAL-3) in IFRS1 cells (Sango et al. 2011b). GAL-3 is a member of a family of β-galactoside-binding animal lectins, and it regulates cell-to-cell and cell-to-matrix interactions. Similar to RAGE, *p60* (AGE-R1), and *p90* (AGE-R2), GAL-3 is identified as an AGE-binding protein (Pricci et al. 2000). Aragno et al. (2005) reported that the upregulation of RAGE and GAL-3 in the hippocampus of STZ-diabetic rats was inhibited by treatment with antioxidants. In addition, it is of interest that exogenous GAL-3 inhibits the proliferation of Schwann cells in cultured sciatic nerve segments (Gustavsson et al. 2007). Further studies are required to determine whether the upregulation of GAL-3 in Schwann cells under diabetic conditions is involved in the pathogenesis of diabetic neuropathy, including glycation, oxidative stress, and reduced regenerative capability.

10.6 Conclusion

We have established immortalized Schwann cells from normal adult mice (IMS32) and rats (IFRS1) as well as murine models of inherited diseases. Because of the biological properties of mature Schwann cells and their high proliferative activities, these Schwann cell lines can be valuable tools for exploring neuron–Schwann cell interactions, the pathobiology of axonal degeneration and regeneration in the PNS, and novel therapeutic approaches against neurological disorders in patients with relevant diseases.

Acknowledgments The work of our laboratory reported in this review was supported by a Grant-in-aid for Scientific Research from the Ministry of Education, Science, Sports and Culture of Japan (grant number: 22500324), the Umehara Fund of the Yokohama Foundation for the

Advancement of Medical Science, Japan, and grants from the Sanwa Kagaku Kenkyusho, Suzuken Memorial Foundation, and the Japan Diabetes Foundation. We thank Drs. Koichi Kato, Yasushi Kanazawa, Shizuka Takaku, Hiroko Yanagisawa, and Miwa Sango-Hirade for helpful suggestions; Emiko Kawakami, Kentaro Endo, and the late Kyoko Ajiki for technical assistance with our studies; Enago (www.enago.jp) for the English language review; and John Wiley and Sons for permission to reproduce the illustrations.

References

Abe K, Namikawa K, Honma M, Iwata T, Matsuoka I, Watabe K, Kiyama H (2001) Inhibition of Ras extracellular-signal-regulated kinase (ERK) mediated signaling promotes ciliary neurotrophic factor (CNTF) expression in Schwann cells. J Neurochem 77(2):700–703

Adler R, Landa KB, Manthorpe M, Varon S (1979) Cholinergic neuronotrophic factors: intraocular distribution of soluble trophic activity for ciliary neurons. Science 204(4400):1434–1436

Airaksinen MS, Saarma M (2002) The GDNF family: signalling, biological functions and therapeutic value. Nat Rev Neurosci 3(5):383–394

Aragno M, Mastrocola R, Medana C, Restivo F, Catalano MG, Pons N, Danni O, Boccuzzi G (2005) Up-regulation of advanced glycated products receptors in the brain of diabetic rats is prevented by antioxidant treatment. Endocrinology 146(12):5561–5567

Arroyo EJ, Scherer SS (2007) The molecular organization of myelinating Schwann cells. In: Armati P (ed) The biology of Schwann cells. Cambridge University Press, New York, pp 37–54

Banerjee TK (2004) Fabry disease with special reference to neurological manifestations. Eur Rev Med Pharmacol Sci 8:275–281

Bolin LM, Iismaa TP, Shooter EM (1992) Isolation of activated adult Schwann cells and a spontaneously immortal Schwann cell clone. J Neurosci Res 33(2):231–238

Brannan CI, Perkins AS, Vogel KS, Ratner N, Nordlund ML, Reid SW, Buchberg AM, Jenkins NA, Parada LF, Copeland NG (1994) Targeted disruption of the neurofibromatosis type-1 gene leads to developmental abnormalities in heart and various neural crest-derived tissues. Genes Dev 8(9):1019–1029

Brockes JP, Fryxell KJ, Lemke GE (1981) Studies on cultured Schwann cells: the induction of myelin synthesis, and the control of their proliferation by a new growth factor. J Exp Biol 95:215–230

Bunge RP (1993) Expanding roles for the Schwann cell: ensheathment, myelination, trophism and regeneration. Curr Opin Neurobiol 3(5):805–809

Calcutt NA, Allendoerfer KL, Mizisin AP, Middlemas A, Freshwater JD, Burgers M, Ranciato R, Delcroix JD, Taylor FR, Shapiro R, Strauch K, Dudek H, Engber TM, Galdes A, Rubin LL, Tomlinson DR (2003) Therapeutic efficacy of sonic hedgehog protein in experimental diabetic neuropathy. J Clin Invest 111(4):507–514

Camby I, Le Mercier M, Lefranc F, Kiss R (2006) Galectin-1: a small protein with major functions. Glycobiology 16(11):137R–157R

Carstea ED, Morris JA, Coleman KG, Loftus SK, Zhang D, Cummings C, Gu J, Rosenfeld MA, Pavan WJ, Krizman DB, Nagle J, Polymeropoulos MH, Sturley SL, Ioannou YA, Higgins ME, Comly M, Cooney A, Brown A, Kaneski CR, Blanchette-Mackie EJ, Dwyer NK, Neufeld EB, Chang TY, Liscum L, Strauss JF 3rd, Ohno K, Zeigler M, Carmi R, Sokol J, Markie D, O'Neill RR, van Diggelen OP, Elleder M, Patterson MC, Brady RO, Vanier MT, Pentchev PG, Tagle DA (1997) Niemann–Pick C1 disease gene: homology to mediators of cholesterol homeostasis. Science 277(5323):228–231

Chi H, Horie H, Hikawa N, Takenaka T (1993) Isolation and age-related characterization of mouse Schwann cells from dorsal root ganglion explants in type I collagen gels. J Neurosci Res 35(2):183–187

Cichowski K, Jacks T (2001) NF1 tumor suppressor gene function: narrowing the GAP. Cell 104(4):593–604

Cooper DN, Barondes SH (1990) Evidence for export of a muscle lectin from cytosol to extracellular matrix and for a novel secretory mechanism. J Cell Biol 110(5):1681–1691

De Vries GH, Boullerne AI (2010) Glial cell lines: an overview. Neurochem Res 35(12):1978–2000

Dyck PJ, Giannini C (1996) Pathologic alterations in the diabetic neuropathies of humans: a review. J Neuropathol Exp Neurol 55(12):1181–1193

Eccleston PA, Mirsky R, Jessen KR (1991) Spontaneous immortalisation of Schwann cells in culture: short-term cultured Schwann cells secrete growth inhibitory activity. Development (Camb) 112(1):33–42

Eckersley L (2002) Role of the Schwann cell in diabetic neuropathy. Int Rev Neurobiol 50:293–321

Fleming CE, Mar FM, Franquinho F, Sousa MM (2009) Transthyretin: an enhancer of nerve regeneration. Int Rev Neurobiol 87:337–346

Friedman B, Scherer SS, Rudge JS, Helgren M, Morrisey D, McClain J, Wang D, Wiegand SJ, Furth ME, Lindsay RM, Ip NY (1992) Regulation of ciliary neurotrophic factor expression in myelin-related Schwann cells in vivo. Neuron 9(2):295–305

Fukunaga M, Miyata S, Liu BF, Miyazaki H, Hirota Y, Higo S, Hamada Y, Ueyama S, Kasuga M (2004) Methylglyoxal induces apoptosis through activation of p38 MAPK in rat Schwann cells. Biochem Biophys Res Commun 320(3):689–695

Giese KP, Martini R, Lemke G, Soriano P, Schachner M (1992) Mouse P0 gene disruption leads to hypomyelination, abnormal expression of recognition molecules, and degeneration of myelin and axons. Cell 71(4):565–576

Gingras M, Beaulieu MM, Gagnon V, Durham HD, Berthod F (2008) In vitro study of axonal migration and myelination of motor neurons in a three-dimensional tissue-engineered model. Glia 56:354–364

Gustavsson P, Linsmeier CE, Leffler H, Kanje M (2007) Galectin-3 inhibits Schwann cell proliferation in cultured sciatic nerve. Neuroreport 18(7):669–673

Haastert K, Mauritz C, Chaturvedi S, Grothe C (2007) Human and rat adult Schwann cell cultures: fast and efficient enrichment and highly effective non-viral transfection protocol. Nat Protoc 2(1):99–104

Hashimoto M, Ishii K, Nakamura Y, Watabe K, Kohsaka S, Akazawa C (2008) Neuroprotective effect of sonic hedgehog up-regulated in Schwann cells following sciatic nerve injury. J Neurochem 107(4):918–927

Horie H, Inagaki Y, Sohma Y, Nozawa R, Okawa K, Hasegawa M, Muramatsu N, Kawano H, Horie M, Koyama H, Sakai I, Takeshita K, Kowada Y, Takano M, Kadoya T (1999) Galectin-1 regulates initial axonal growth in peripheral nerves after axotomy. J Neurosci 19(22):9964–9974

Horie H, Kadoya T, Hikawa N, Sango K, Inoue H, Takeshita K, Asawa R, Hiroi T, Sato M, Yoshioka T, Ishikawa Y (2004) Oxidized galectin-1 stimulates macrophages to promote axonal regeneration in peripheral nerves after axotomy. J Neurosci 24(8):1873–1880

Ito Y, Wiese S, Funk N, Chittka A, Rossoll W, Bömmel H, Watabe K, Wegner M, Sendtner M (2006) Sox10 regulates ciliary neurotrophic factor gene expression in Schwann cells. Proc Natl Acad Sci USA 103(20):7871–7876

Kawashima I, Watabe K, Tajima Y, Fukushige T, Kanzaki T, Kanekura T, Sugawara K, Ohyanagi N, Suzuki T, Togawa T, Sakuraba H (2007) Establishment of immortalized Schwann cells from Fabry mice and their low uptake of recombinant alpha-galactosidase. J Hum Genet 52(12):1018–1025

Kieseier BC, Hu W, Hartung H-P (2007) Schwann cells as immunomodulatory cells. In: Armati P (ed) The biology of Schwann cells. Cambridge University Press, New York, pp 118–125

Kleitman N, Wood PM, Bunge RP (1998) Tissue culture methods for the study of myelination. In: Banker G, Goslin K (eds) Culturing nerve cells, 2nd edn. MIT Press, Cambridge, pp 545–594

Kobayashi T, Yamanaka T, Jacobs JM, Teixeira F, Suzuki K (1980) The Twitcher mouse: an enzymatically authentic model of human globoid cell leukodystrophy (Krabbe disease). Brain Res 202(2):479–483

Lehmann HC, Höke A (2010) Schwann cells as a therapeutic target for peripheral neuropathies. CNS Neurol Disord Drug Targets 9(6):801–806

Lehmann HC, Köhne A, Bernal F, Jangouk P, Meyer Zu Hörste G, Dehmel T, Hartung HP, Previtali SC, Kieseier BC (2009) Matrix metalloproteinase-2 is involved in myelination of dorsal root ganglia neurons. Glia 57:479–489

Li R (1998) Culture methods for selective growth of normal rat and human Schwann cells. Methods Cell Biol 57:167–186

Loftus SK, Morris JA, Carstea ED, Gu JZ, Cummings C, Brown A, Ellison J, Ohno K, Rosenfeld MA, Tagle DA, Pentchev PG, Pavan WJ (1997) Murine model of Niemann–Pick C disease: mutation in a cholesterol homeostasis gene. Science 277(5323):232–235

Magnaghi V, Procacci P, Tata AM (2009) Novel pharmacological approaches to Schwann cells as neuroprotective agents for peripheral nerve regeneration. Int Rev Neurobiol 87:295–315

Mathon NF, Malcolm DS, Harrisingh MC, Cheng L, Lloyd AC (2001) Lack of replicative senescence in normal rodent glia. Science 291(5505):872–875

Mirsky R, Jessen KR (2007) Early events in Schwann cell development. In: Armati P (ed) The biology of Schwann cells. Cambridge University Press, New York, pp 13–36

Miyawaki S, Mitsuoka S, Sakiyama T, Kitagawa T (1982) Sphingomyelinosis, a new mutation in the mouse: a model of Niemann–Pick disease in humans. J Hered 73(4):257–263

Mizisin AP, Li L, Perello M, Freshwater JD, Kalichman MW, Roux L, Calcutt NA (1996) Polyol pathway and osmoregulation in JS1 Schwann cells grown in hyperglycemic and hyperosmotic conditions. Am J Physiol 270(1 pt 2):F90–F97

Muir D, Varon S, Manthorpe M (1990) Schwann cell proliferation in vitro is under negative autocrine control. J Cell Biol 111(6 pt 1):2663–2671

Murakami T, Ohsawa Y, Zhenghua L, Yamamura K, Sunada Y (2010) The transthyretin gene is expressed in Schwann cells of peripheral nerves. Brain Res 1348:222–225

Naureckiene S, Sleat DE, Lackland H, Fensom A, Vanier MT, Wattiaux R, Jadot M, Lobel P (2000) Identification of HE1 as the second gene of Niemann–Pick C disease. Science 290(5500):2298–2301

Nishikawa T, Araki E (2007) Impact of mitochondrial ROS production in the pathogenesis of diabetes mellitus and its complications. Antioxid Redox Signal 9(3):343–353

Ogata T, Iijima S, Hoshikawa S, Miura T, Yamamoto S, Oda H, Nakamura K, Tanaka S (2004) Opposing extracellular signal-regulated kinase and Akt pathways control Schwann cell myelination. J Neurosci 24(30):6724–6732

Ohsawa M, Kotani M, Tajima Y, Tsuji D, Ishibashi Y, Kuroki A, Itoh K, Watabe K, Sango K, Yamanaka S, Sakuraba H (2005) Establishment of immortalized Schwann cells from Sandhoff mice and corrective effect of recombinant human beta-hexosaminidase A on the accumulated GM2 ganglioside. J Hum Genet 50(9):460–467

Ohshima T, Murray GJ, Swaim WD, Longenecker G, Quirk JM, Cardarelli CO, Sugimoto Y, Pastan I, Gottesman MM, Brady RO, Kulkarni AB (1997) Alpha-galactosidase A deficient mice: a model of Fabry disease. Proc Natl Acad Sci USA 94(6):2540–2544

Ota K, Nakamura J, Li W, Kozakae M, Watarai A, Nakamura N, Yasuda Y, Nakashima E, Naruse K, Watabe K, Kato K, Oiso Y, Hamada Y (2007) Metformin prevents methylglyoxal-induced apoptosis of mouse Schwann cells. Biochem Biophys Res Commun 357(1):270–275

Padilla A, Descorbeth M, Almeyda AL, Payne K, De Leon M (2011) Hyperglycemia magnifies Schwann cell dysfunction and cell death triggered by PA-induced lipotoxicity. Brain Res 1370:64–79

Paratcha G, Ledda F, Ibáñez CF (2003) The neural cell adhesion molecule NCAM is an alternative signaling receptor for GDNF family ligands. Cell 113(7):867–879

Peden KW, Charles C, Sanders L, Tennekoon GI (1989) Isolation of rat Schwann cell lines: use of SV40 T antigen gene regulated by synthetic metallothionein promoters. Exp Cell Res 185(1):60–72

Pentchev PG, Gal AE, Booth AD, Omodeo-Sale F, Fouks J, Neumeyer BA, Quirk JM, Dawson G, Brady RO (1980) A lysosomal storage disorder in mice characterized by a dual deficiency of sphingomyelinase and glucocerebrosidase. Biochim Biophys Acta 619(3):669–679

Plachta N, Annaheim C, Bissière S, Lin S, Rüegg M, Hoving S, Müller D, Poirier F, Bibel M, Barde YA (2007) Identification of a lectin causing the degeneration of neuronal processes using engineered embryonic stem cells. Nat Neurosci 10(6):712–719

Porter S, Glaser L, Bunge RP (1987) Release of autocrine growth factor by primary and immortalized Schwann cells. Proc Natl Acad Sci USA 84(21):7768–7772

Pricci F, Leto G, Amadio L, Iacobini C, Romeo G, Cordone S, Gradini R, Barsotti P, Liu FT, Di Mario U, Pugliese G (2000) Role of galectin-3 as a receptor for advanced glycosylation end products. Kidney Int Suppl 77:S31–S39

Sandhoff K (2001) The GM2 gangliosidoses and the elucidation of the β-hexosaminidase system. In: Desnick RJ, Kaback MM (eds) Tay–Sachs disease, vol 44, Advances in genetics. Academic, San Diego, pp 67–91

Sango K, Verdes JM, Hikawa N, Horie H, Tanaka S, Inoue S, Sotelo JR, Takenaka T (1994) Nerve growth factor (NGF) restores depletions of calcitonin gene-related peptide and substance P in sensory neurons from diabetic mice in vitro. J Neurol Sci 126(1):1–5

Sango K, Yamanaka S, Hoffmann A, Okuda Y, Grinberg A, Westphal H, McDonald MP, Crawley JN, Sandhoff K, Suzuki K, Proia RL (1995) Mouse models of Tay–Sachs and Sandhoff diseases differ in neurologic phenotype and ganglioside metabolism. Nat Genet 11(2):170–176

Sango K, Tokashiki A, Ajiki K, Horie M, Kawano H, Watabe K, Horie H, Kadoya T (2004) Synthesis, localization and externalization of galectin-1 in mature dorsal root ganglion neurons and Schwann cells. Eur J Neurosci 19(1):55–64

Sango K, Saito H, Takano M, Tokashiki A, Inoue S, Horie H (2006a) Cultured adult animal neurons and Schwann cells give us new insights into diabetic neuropathy. Curr Diabetes Rev 2(2):169–183

Sango K, Suzuki T, Yanagisawa H, Takaku S, Hirooka H, Tamura M, Watabe K (2006b) High glucose-induced activation of the polyol pathway and changes of gene expression profiles in immortalized adult mouse Schwann cells IMS32. J Neurochem 98(2):446–458

Sango K, Yanagisawa H, Komuta Y, Si Y, Kawano H (2008a) Neuroprotective properties of ciliary neurotrophic factor for cultured adult rat dorsal root ganglion neurons. Histochem Cell Biol 130(4):669–679

Sango K, Yanagisawa H, Kato K, Kato N, Hirooka H, Watabe K (2008b) Differential effects of high glucose and methylglyoxal on viability and polyol metabolism in immortalized adult mouse Schwann cells. Open Diabetes J 1:1–11

Sango K, Yanagisawa H, Kawakami E, Takaku S, Ajiki K, Watabe K (2011a) Spontaneously immortalized Schwann cells from adult Fischer rat as a valuable tool for exploring neuron-Schwann cell interactions. J Neurosci Res 89(6):898–908

Sango K, Yanagisawa H, Takaku S, Kawakami E, Watabe K (2011b) Immortalized adult rodent Schwann cells as in vitro models to study diabetic neuropathy. Exp Diabetes Res 2011:374943

Sango K, Yanagisawa H, Watabe K, Horie H, Kadoya T (2012a) Galectin-1 as a multifunctional molecule in the peripheral nervous system after injury. In: Rayegani SM (ed) Basic principles of peripheral nerve disorders. InTech Doo, Rijeka, pp 31–46 http://www.intechopen.com/books/basic-principles-of-peripheral-nerve-disorders/galectin-1-as-a-multifunctional-molecule-in-the-peripheral-nervous-system-after-injury

Sango K, Kawakami E, Yanagisawa H, Takaku S, Tsukamoto M, Utsunomiya K, Watabe K (2012b) Myelination in coculture of established neuronal and Schwann cell lines. Histochem Cell Biol 137(6):829–839

Sasaki T, Hirabayashi J, Manya H, Kasai K, Endo T (2004) Galectin-1 induces astrocyte differentiation, which leads to production of brain-derived neurotrophic factor. Glycobiology 14(4):357–363

Sendtner M, Stockli KA, Thoenen H (1992) Synthesis and localization of ciliary neurotrophic factor in the sciatic nerve of the adult rat after lesion and during regeneration. J Cell Biol 118(1):139–148

Shen JS, Watabe K, Meng XL, Ida H, Ohashi T, Eto Y (2002) Establishment and characterization of spontaneously immortalized Schwann cells from murine model of globoid cell leukodystrophy (twitcher). J Neurosci Res 68(5):588–594

Song Z, Fu DT, Chan YS, Leung S, Chung SS, Chung SK (2003) Transgenic mice overexpressing aldose reductase in Schwann cells show more severe nerve conduction velocity deficit and oxidative stress under hyperglycemic stress. Mol Cell Neurosci 23(4):638–647

Sousa MM, Saraiva MJ (2003) Neurodegeneration in familial amyloid polyneuropathy: from pathology to molecular signaling. Prog Neurobiol 71(5):385–400

Sugimoto K, Yasujima M, Yagihashi S (2008) Role of advanced glycation end products in diabetic neuropathy. Curr Pharm Des 14(10):953–961

Suzuki K, Vanier MT, Suzuki K (1999) Lysosomal disorders. In: Popko B (ed) Mouse models in the study of genetic neurological disorders, vol 9, Advances in neurochemistry. Kluwer, New York, pp 245–283

Suzuki J, Akahane K, Nakamura J, Naruse K, Kamiya H, Himeno T, Nakamura N, Shibata T, Kondo M, Nagasaki H, Fujiya A, Oiso Y, Hamada Y (2011) Palmitate induces apoptosis in Schwann cells via both ceramide-dependent and independent pathways. Neuroscience 176:188–198

Takaku S, Yanagisawa H, Watabe K, Horie H, Kadoya T, Sakumi K, Nakabeppu Y, Poirier F, Sango K (2013) GDNF promotes neurite outgrowth and upregulates galectin-1 through the RET/PI3K signaling in cultured adult rat dorsal root ganglion neurons. Neurochem Int 62(3):330–339

Toda K, Small JA, Goda S, Quarles RH (1994) Biochemical and cellular properties of three immortalized Schwann cell lines expressing different levels of the myelin-associated glycoprotein. J Neurochem 63(5):1646–1657

Tosaki T, Kamiya H, Yasuda Y, Naruse K, Kato K, Kozakae M, Nakamura N, Shibata T, Hamada Y, Nakashima E, Oiso Y, Nakamura J (2008) Reduced NGF secretion by Schwann cells under the high glucose condition decreases neurite outgrowth of DRG neurons. Exp Neurol 213(2):381–387

Traiffort E, Angot E, Ruat M (2010) Sonic Hedgehog signaling in the mammalian brain. J Neurochem 113(3):576–590

Verdú E, Ceballos D, Vilches JJ, Navarro X (2000) Influence of aging on peripheral nerve function and regeneration. J Peripher Nerv Syst 5(4):191–208

Watabe K, Yamada M, Kawamura T, Kim SU (1990) Transfection and stable transformation of adult mouse Schwann cells with SV-40 large T antigen gene. J Neuropathol Exp Neurol 49(5):455–467

Watabe K, Fukuda T, Tanaka J, Toyohara K, Sakai O (1994) Mitogenic effects of platelet-derived growth factor, fibroblast growth factor, transforming growth factor-beta, and heparin-binding serum factor for adult mouse Schwann cells. J Neurosci Res 39(5):525–534

Watabe K, Fukuda T, Tanaka J, Honda H, Toyohara K, Sakai O (1995) Spontaneously immortalized adult mouse Schwann cells secrete autocrine and paracrine growth-promoting activities. J Neurosci Res 41(2):279–290

Watabe K, Ida H, Uehara K, Oyanagi K, Sakamoto T, Tanaka J, Garver WS, Miyawaki S, Ohno K, Eto Y (2001) Establishment and characterization of immortalized Schwann cells from murine model of Niemann–Pick disease type C (spm/spm). J Peripher Nerv Syst 6(2):85–94

Watabe K, Sakamoto T, Kawazoe Y, Michikawa M, Miyamoto K, Yamamura T, Saya H, Araki N (2003) Tissue culture methods to study neurological disorders: establishment of immortalized Schwann cells from murine disease models. Neuropathology 23(1):68–78

Yamauchi J, Miyamoto Y, Chan JR, Tanoue A (2008) ErbB2 directly activates the exchange factor Dock7 to promote Schwann cell migration. J Cell Biol 181(2):351–365

Zhang L, Ma Z, Smith GM, Wen X, Pressman Y, Wood PM, Xu XM (2009) GDNF-enhanced axonal regeneration and myelination following spinal cord injury is mediated by primary effects on neurons. Glia 57:1178–1191

Index

A
Activating transcription factor, 57
Adherens and tight junctions, 32
Advanced glycation end product (AGE), 156, 165
Akt, 19
Aldose reductase (AR), 155
Amyloid polyneuropathy, 3
aPKC, 10
Arf6, 13
Arfgef10, 21
Arp2/3, 17
Ascorbic acid, 96, 161, 164
Astrocyte, 16
Autophagy, 95
Autopsy cases, 111
Axon, 39
Axonal degeneration, 71
Axonal mitochondria, 2
Axonal outgrowth, 48

B
β-amyloid converting enzyme (BACE1), 10, 52
Bands of Büngner, 75
Basal lamina, 52
Brain-derived neurotrophin factor (BDNF), 13, 48
B27 supplement, 161, 163

C
CADM1, 34
CADM4. *See* Cell adhesion molecule 4 (CADM4)
Calpain, 127
Caspr2, 51
β-Catenin, 35, 51
Cdc42, 10
Cell adhesion molecule 4 (CADM4), 35
Cell elongation, 6
Cell migration, 6
Charcot-Marie-Tooth diseases, 3, 53
Charcot-Marie-Tooth disease type 1B (CMT1B), 157, 159
Chromatin remodeling, 6
Chronic inflammatory demyelinating polyradiculoneuropathy (CIDP), 128
Ciliary neurotrophic factor (CNTF), 152, 153
c-Jun, 73
Co-culture, 160
Complement, 125
Connexin 32, 59
Contactin, 50
Contactin associated protein, 50
C-peptide, 143
Crush injury, 71
Cyclic AMP, 10
Cytohesin-1, 13

D
Dbs, 13
Demyelination, 2, 56, 128
Diabetic neuropathy, 3, 136
Discs large 1 (Dlg1), 36
Dock7, 10
Dominant-negative function, 88
Dorsal root ganglia/ganglion (DRG), 13, 104, 150
  neurons, 160, 164
  sensory neurons, 114
Dysmyelination, 56

K. Sango and J. Yamauchi (eds.), *Schwann Cell Development and Pathology*,
DOI 10.1007/978-4-431-54764-8, © Springer Japan 2014

E

E-cadherin, 35
EphA, 16
EphB, 16
Ephrin, 16
Epithelial cells, 33
ErbB, 52
ErbB2, 10, 75
ErbB3, 10, 75, 76
ER stress, 86, 87, 94
Erythrocyte, 30
Erythropoietin, 15
E3 ubiquitin ligase, 87
Ezrin-radixin-moesin (ERM) domain, 31

F

Fabry disease, 159
Familial amyloidotic polyneuropathy (FAP), 104
FGD4, 20
Forskolin, 150
Fyn, 12

G

4.1G, 31
β-Galactosidase, 162
Galectin-1 (GAL-1), 152, 154, 164
Galectin-3, 165
Ganglioside, 125, 127
Ganglioside induced differentiation associated protein 1, 57
Gap junction, 89
Gas6, 14
GDNF. *See* Glial cell line-derived neurotrophic factor (GDNF)
Genetic testing, 84, 85
Glial cell line-derived neurotrophic factor (GDNF), 14, 152, 161, 164
Gliomedin, 49, 129
Glycation, 141–142
Glyceraldehyde-3-phosphate dehydrogenase, 58
Glycolysis, 58
GPR126, 19
GSK3, 19
GTPase-activating protein (GAP), 10
Guanine-nucleotide exchange factor (GEF), 10
Guanylate kinase (MAGUK), 31
Guillain–Barré syndrome (GBS), 124

H

Human TTR Met30 gene, 108

I

IFRS1, 3, 150, 160
IGF1, 13
IGF1R, 14
Immortalized mouse Schwann cells (IMS32 cells), 137
Immune-mediated neuropathies, 3, 122
IMS32, 3, 108, 150, 151
In vivo cryotechnique, 31

J

JNK, 10
Juxtaparanode, 51

K

K^+ channels, 50
Kinase, 10
Krabbe disease, 157, 158

L

L1, 51
Lactate, 58
LacZ-labeled IFRS1, 162
Laser capture microdissection (LCM), 106
LIMK, 17
Linkage analyses, 84
LRP1, 15

M

Mature Schwann cell, 5
Maturity, 40
Mechanical strength, 35
Membrane-associated, 31
Membrane palmitoylated protein 6 (MPP6), 31
Membrane skeleton, 30
Membrane stability, 39
Membrane trafficking, 91, 92, 97
Membrane transport, 92
Merlin, 34
Methylglyoxal (MG), 156
Misfolding, 87, 93
Mitochondria, 55, 57
Mitochondrial function, 89
Mitochondrial transcription factor A, 59
Mitofusin 2, 57
Molecular complex, 40
Monocarboxylic acid transporter 1, 58
Morphofunctional mechanism, 41
MPP6, 32
mRNA, 6

Mutant mice, 85, 90
Mycobacterium leprae, 21
Myelin, 70
Myelin-associated glycoprotein, 54
Myelinated nerve fibers, 40
Myelinating Schwann cell, 6
Myelination, 1, 6, 40, 160
Myelin protein zero (MPZ or P0), 54, 159
Myelin sheath, 5, 31

N
NAB, 88
Na^+ channels, 49, 55
Na^+/K^+-ATPases, 55
Nanofiber, 21
N-cadherin, 51
N-CAM, 51
Nectin-like protein, 51, 76
Nerve biopsies, 83
Nerve conduction block, 71
Nerve conduction velocity (NCV), 83
Nerve growth factor (NGF), 48, 152, 156, 163
Neural cell adhesion molecule (NCAM), 14
Neural crest cell, 5
Neuregulin-1, 75
Neuregulin (NRG)-β, 150
Neuregulin-1 isotype SMDF, 163
Neuregulins, 52
Neurofascin 155, 50
Neurofascin 186, 49
Neurofibromatosis, 157
Neurofibromatosis type I (NF1), 160
Neurotoxicity, 112–113
Neurotrophin, 53
Neurotrophin secretion, 142
NF186, 129
NGF. *See* Nerve growth factor (NGF)
Niemann–Pick disease type C (NPC), 157
Nodes of Ranvier, 3, 49, 70, 122, 123
Nonfibrillar aggregates, 112
Nonmyelinating Schwann cell, 6
NrCAM, 49
NRG1, 9
NRG1 type III, 10
N2 supplement, 163
NT3, 13
N-WASP, 17

O
Oncogenic kinase, 36
Onion bulb shape, 83
Oxidative stress, 139–141

P
p21-activated kinase (PAK), 17
Par3, 10
Par6, 10
Paranodes, 49
Pathogenesis, 136
Pathological condition, 41
PC12, 163
p120 catenin, 39
PC12 cells, 161
PDZ-domain, 91
Peripheral myelin protein 22, 54
Peripheral nerves, 107
Perisynaptic Schwann cell, 6
Phosphorylation/dephosphorylation, 37
PI3K, 14
PKA, 19
p57kip2, 16
Platelet-derived growth factor (PDGF), 19
Plexin, 15
$p75^{NTR}$, 10, 13
Point mutations, 86
Polyneuropathy, 82
Polyol pathway, 138, 155
Premyelinating process, 6
Progesterone, 96
Proteasome system(s), 86, 94
Protein 4.1B, 50
Protein kinase C, 139
PSD (postsynaptic density)-95/Dlg (Drosophila disks large)/ZO (zonula occludens)-1 (PDZ), 32

R
Rac1, 10, 40
Rap, 19
Ras extracellular signal-regulated kinase (ERK), 154
Reactive oxygen species (ROS), 156
Recycling, 90
Regeneration, 71
Remak bundles, 51
"Remak" fibers, 48
Remyelination, 71, 75
Resveratrol, 140
Reverse transcription-polymerase chain reaction (RT-PCR), 106
RhoA, 13
Rho-kinase, 17
RNA interference, 96
RT4, 108

S

Sandhoff disease, 157, 158
Satellite cell(s), 6, 106
Schmidt–Lanterman incisure, 2, 31
Schwann cell(s), 5, 106
  abnormalities, 111
  hypothesis, 113
  lineage, 5
  precursor, 5
Schwannoma, 39, 148
Semaphorin, 15
Septate-like junctions, 49–50
Shape, 35
Shp2, 11
Small GTPase, 10
Sonic Hedgehog (Shh), 154
Sox10, 154
Spectrin, 76
Spectrin–actin network, 30
Spontaneous immortalization, 148
Src kinase family, 37
Sudan black B staining, 162
Sural nerve biopsies, 110
SV40 large T-antigen, 148

T

TRAF6, 13
Transcription factor, 6
Transdifferentiation, 75
Transient axonal glycoprotein-1, 51
Transthyretin (TTR), 104, 157
TrkC, 13
TTR amyloid deposits, 110
Tumor necrosis factor-alpha (α)-converting enzyme (TACE), 10, 53
Twitcher mouse, 158
Tyro3, 14

U

Ubiquitination, 86

V

Vav2, 13
Voltage-gated Na^+ (Nav) channel, 123
von Recklinghausen neurofibromatosis, 160

W

Wallerian degeneration, 40

Z

Zebrafish, 8

MIX
Papier aus verantwortungsvollen Quellen
Paper from responsible sources
FSC® C105338

If you have any concerns about our products,
you can contact us on
ProductSafety@springernature.com

In case Publisher is established outside the EU,
the EU authorized representative is:
Springer Nature Customer Service Center GmbH
Europaplatz 3, 69115 Heidelberg, Germany

Printed by Libri Plureos GmbH
in Hamburg, Germany